PRAISE FOR *THE FLY IN THE OINTMENT*

"Joe Schwarcz has done it again. In fact, h
every bit as entertaining, informative, and authoritative as his previous celebrated collections, but contains enriched social fiber and 10 percent more attitude per chapter. Whether he's assessing the legacy of Rachel Carson, coping with penile underachievement in alligators, or revealing the curdling secrets of cheese, Schwarcz never fails to fascinate."

— Curt Supplee, former science editor, *Washington Post*

"Wanna know how to wow 'em at a cocktail party or in a chemistry classroom? Take a stroll through the peripatetic journalistic world of the ideas and things of science brought to life by Dr. Joe. Here is narrative science at its best. The end result? In either place, scientific literacy made useful by *The Fly in the Ointment*."

— Leonard Fine, professor of chemistry, Columbia University

PRAISE FOR *DR. JOE AND WHAT YOU DIDN'T KNOW*

"Any science writer can come up with the answers. But only Dr. Joe can turn the world's most fascinating questions into a compelling journey through the great scientific mysteries of everyday life. *Dr. Joe and What You Didn't Know* proves yet again that all great science springs from the curiosity of asking the simple question . . . and that Dr. Joe is one of the great science storytellers with both all the questions and answers."

— Paul Lewis, president and general manager, Discovery Channel

PRAISE FOR *THAT'S THE WAY THE COOKIE CRUMBLES*

"Schwarcz explains science in such a calm, compelling manner, you can't help but heed his words. How else to explain why I'm now stir-frying cabbage for dinner and seeing its cruciferous cousins — broccoli, cauliflower, and brussels sprouts — in a delicious new light?"

— Cynthia David, *Toronto Star*

PRAISE FOR *RADAR, HULA HOOPS, AND PLAYFUL PIGS*

"It is hard to believe that anyone could be drawn to such a dull and smelly subject as chemistry — until, that is, one picks up Joe Schwarcz's book and is reminded that with every breath and feeling one is experiencing chemistry. Falling in love, we all know, is a matter of the right chemistry. Schwarcz gets his chemistry right, and hooks his readers."

— John C. Polanyi, Nobel Laureate

ALSO BY DR. JOE SCHWARCZ

Is That a Fact?: Frauds, Quacks, and the Real Science of Everyday Life

The Right Chemistry: 108 Enlightening, Nutritious, Health-Conscious, and Occasionally Bizarre Inquiries into the Science of Everyday Life

Dr. Joe's Health Lab: 164 Amazing Insights into the Science of Medicine, Nutrition, and Well-Being

Dr. Joe's Brain Sparks: 179 Inspiring and Enlightening Inquiries into the Science of Everyday Life

Dr. Joe's Science, Sense & Nonsense: 61 Nourishing, Healthy, Bunk-Free Commentaries on the Chemistry That Affects Us All

Brain Fuel: 199 Mind-Expanding Inquiries into the Science of Everyday Life

An Apple a Day: The Myths, Misconceptions, and Truths About the Foods We Eat

Let Them Eat Flax: 70 All-New Commentaries on the Science of Everyday Food & Life

The Fly in the Ointment: 70 Fascinating Commentaries on the Science of Everyday Life

Dr. Joe and What You Didn't Know: 177 Fascinating Questions and Answers About the Chemistry of Everyday Life

That's the Way the Cookie Crumbles: 62 All-New Commentaries on the Fascinating Chemistry of Everyday Life

The Genie in the Bottle: 64 All-New Commentaries on the Fascinating Chemistry of Everyday Life

Radar, Hula Hoops, and Playful Pigs: 67 Digestible Commentaries on the Fascinating Chemistry of Everyday Life

A FEAST OF SCIENCE

Intriguing Morsels from the Science of Everyday Life

Dr. JOE SCHWARCZ

Published in Canada by ECW Press
665 Gerrard Street East
Toronto, Ontario, Canada M4M 1Y2
416-694-3348 / info@ecwpress.com

Cover design: David A. Gee

LIBRARY AND ARCHIVES CANADA
CATALOGUING IN PUBLICATION

Schwarcz, Joe, author
A feast of science : intriguing morsels from the science of everyday life / Dr. Joe Schwarcz.

Includes index.
Issued in print and electronic formats.
ISBN 978-1-77041-192-0 (softcover).
ALSO ISSUED AS: 978-1-77305-135-2 (PDF),
978-1-77305-134-5 (ePUB)

1. Science—Popular works. 2. Science—Miscellanea.

I. TITLE.

Q173.S3845 2018 500 C2017-906589-0
C2017-906590-4

The publication of *A Feast of Science* has been generously supported by the Government of Canada through the Canada Book Fund. *Ce livre est financé en partie par le gouvernement du Canada*. We also acknowledge the support of the Government of Ontario through the Ontario Book Publishing Tax Credit and the Ontario Media Development Corporation.

Canada

PRINTED AND BOUND IN CANADA

PRINTING: FRIESENS 5 4 3 2 1

INTRODUCTION

I feast on science. I love biting into a juicy piece of research, and I find morsels of science history to be very tasty. Fortunately, these days there is a veritable banquet of succulent science servings available; one just has to know at which table to sit. Unfortunately, sitting down at the wrong one can lead to a science famine instead of a feast. That's because the Internet and the various forms of social media serve up a steady stream of dishes that are as likely to be filled with toxic misinformation as with delectable science. So, how do we differentiate between the two? It is a question of looking for evidence. That, though, can be quite a challenge.

When we take a medication, we trust that there is evidence that it will work. When we apply a cosmetic, we trust there is evidence that it is safe. When we put on a sunscreen, we trust that there is evidence that it filters ultraviolet light. But evidence is not white or black; it runs the gamut from anecdotal to incontrovertible.

Some people claim that placing a bar of soap under the sheet before they sleep cures restless leg syndrome. That's what we call anecdotal evidence, and it remains so until it is confirmed or dismissed by proper randomized double-blind controlled trials. On the other hand, evidence that gold conducts electricity is ironclad. There are no ifs or buts about it. Often, though, the use of the term "evidence" is open to interpretation. An interesting example is a paper published in *Nature*, one of the world's foremost scientific journals, intriguingly titled: "Evidence for Human Transmission of Amyloid-[Beta] Pathology and Cerebral Amyloid Angiopathy."

Let's dissect this title. Amyloid-beta proteins are one of the hallmarks of Alzheimer's disease, so the title implies that evidence has been found that the disease can be transmitted from person to person. Little wonder that the paper generated headlines in the lay press ranging from "Alzheimer's May Be a Transmissible Infection" and "You Can Catch Alzheimer's" to "Alzheimer's Bombshell." All of these are highly misleading because the paper, in spite of its provocative title, does not provide evidence for the transmission of Alzheimer's disease between humans.

So what did the researchers, led by Dr. John Collinge of University College London, actually find? They investigated the brains of eight people who had been injected with human growth hormone as children due to stunted growth back when this hormone was extracted from the pituitary glands of dead donors. Unfortunately, the donors from whom the hormone was extracted had been harboring proteins known as prions that cause Creutzfeldt-Jakob disease, a terminal neurological affliction. The recipients ended up dying from the disease they had contracted via the hormone.

Collinge found that six of the eight people also had amyloid plaques typical of Alzheimer's disease. But none of 116 people who had died of Creutzfeldt-Jakob disease and not received contaminated growth hormone showed any sign of amyloid protein deposits. Dr. Collinge therefore suggested that molecules that lead to amyloid plaque formation were passed to the recipients along with the growth hormone.

A very interesting hypothesis to be sure. But the study did not show that the patients would actually have developed Alzheimer's had they lived longer, or that the growth hormone was contaminated with molecules that can "seed" Alzheimer's disease. A more appropriate title for the paper would have been "Possibility for human transmission of amyloid-beta pathology via contaminated growth hormone." The word "evidence," which in this case is at best circumstantial, should not have appeared in the title. The authors do clearly point out that "there is no suggestion that Alzheimer's disease is a contagious disease and no supportive evidence from epidemiological studies that Alzheimer's disease is transmissible." Nevertheless it was the term "evidence" that caught journalists' eyes and created undue public alarm with the suggestion that Alzheimer's disease can be "caught." The fact is that no such evidence exists.

Those of us who try to be responsible in communicating science and offer advice about its interpretation face a number of challenges. First, we can't offer what the public often yearns for — namely, simple solutions to complex problems. Contrary to what the web may churn out, we don't claim that a cancer cure has been found in blue scorpion venom or that people can be protected from the effects of herbicide residues with a homeopathic "detox." Neither do we propose to use quartz crystals to create harmony between the body and the soul by opening energy channels through which positive energy can cruise. Furthermore, our accounts are often peppered with "ifs," "buts," and "maybes," along with calls for further research.

Second, we have to cope with the view among much of the public that scientists can't be trusted because they are in the pockets of industry. For example, we have to confront the daftness that natural cancer cures are being hidden by "Big Pharma" so it can profit from selling its expensive but ineffective drugs.

Finally, there is the problem that many scientists, who may indeed be outstanding researchers, are quite inept when it comes to communicating their work to the public, leaving laypeople confused and frustrated. Unfortunately, when this happens, the peddlers of seductive nonsense rush in with their miraculous solutions, sometimes even distracting the desperate from remedies that may actually work.

Recently another challenge has surfaced. Thanks to President Trump, terms such as "alternative facts," "fake news," "hoax," "dishonest media," and "believe me" have entered the political lexicon. The inane concept of "alternative facts" is easy to dismiss. There are no such things.

Consider Aristotle's and Galileo's "alternative facts" about two objects of unequal mass being dropped from a height. Aristotle claimed that the heavier object would fall faster. That became a "fact" simply because the respected Aristotle said, "believe me." This "fact" was unchallenged until Galileo proposed that the two objects would hit the ground at the same time and proved it by dropping two cannonballs of unequal mass from the Tower of Pisa, or so the story goes. Any doubt about the actual fact was dispelled when astronaut David Scott dropped a hammer and a feather on the moon, demonstrating that when air resistance is eliminated, objects fall at a rate independent of their mass.

Aristotle's proposal that babies are formed from menstrual blood was also deemed to be a "fact" because it meshed with the observation that menstruation ceases at the onset of pregnancy. Of course, a baby is not made from menstrual blood. It is the secretion of human chorionic gonadotropin (HCG) by the placenta that prevents the uterine lining from being shed. Aristotle's "fact" was not a fact at all.

These days, especially when it comes to health matters, the public is often beguiled by "alternatives" that are presented as facts. Alkaline diets are claimed to treat cancer, vaccines are linked to autism, and aluminum is said to cause Alzheimer's disease. These are about as factual as the "alternative facts" that were dredged up to support President Trump's contention that the crowd at his inauguration was far larger than the one at President Obama's.

What about "fake news" when it comes to science? It certainly exists. A widely circulating news report about Intelligex, a supposed "smart pill," gives the appearance of being on *Forbes* magazine's website. It also features a screenshot from CNN with a picture of physicist Stephen Hawking, who is said to be using the pill to "triple his memory." Of course Hawking never called Intelligex "Viagra of the brain" or predicted the pill would change humanity. He had probably never even heard of Intelligex. The *Forbes* logo, the CNN screenshot, and Hawking's quotes are all fake. But the money people spend on a useless product is real.

If you want an example of a hoax, look no further than the authentic-sounding Health Sciences Institute. It warns of a "DEADLY CRISIS that is sweeping America, killing seniors at a relentless pace with FULL knowledge and APPROVAL of the government." What is this scourge? Prescription drugs! Their seductive spiel goes on to tantalize with "natural" cures for diabetes, arthritis, heart disease, and a substance that "vaporizes cancer in six weeks." To find out what these secret remedies are, all you have to do is join the institute so that you can get a copy of *Miracles from the Vault: Anthology of Underground Cures*. What you'll get is a collection of "alternative facts" not backed by evidence. And when it comes to science, evidence is everything.

So let's start serving up our feast of science and partake of the many dishes it has to offer. We will make it a buffet, so you can wander about, picking up whatever you may deem to be a tasty morsel. But let's start with a little personal history.

A FEAST OF SCIENCE

INFORMATION AND MISINFORMATION

The first time I had a chance to watch television was in 1956 after coming to Canada following the Hungarian Revolution. Back then there was only one channel, and it was on the air for only a few hours a day. The newscasts did provide a window to the world that I had not seen open before. Telephones were already well established, but a call to Europe had to be prearranged. For "breaking news," you depended on local radio stations where you could also tune in to a variety of talk shows. There was the popular Joe Pyne, who would invite you to gargle with razor blades if you disagreed with him, and my favorite, Pat Burns, who had an opinion on everything and was not averse to abusing his callers. Indeed, it was Burns who stimulated my interest in skepticism.

One of the regular callers on the *Burns Hot Line* was a lady who was convinced that space aliens walked among us, specifically, on Montreal's St. Catherine Street. She recognized them because of their distinctive eyes! Pat would humor her for comic relief and often goaded her into making outrageous comments. One day, however, he was stressed for time and told her that he couldn't let her go on about her "little green men." She didn't take this well and claimed that if Pat cut her off, the aliens would cut him off. "OK, tell me tomorrow why they didn't," he retorted, as he proceeded to cut her off. Then he went to the next call, but there wasn't one. The station went off the air and stayed off for six hours. There was no explanation.

The lady called back the next day to gloat, but Pat just said "coincidence, Doll, coincidence." She stuck by her guns and maintained the aliens had swung into action. "So let's see them do it again," Burns fumed as he again cut her off. Well, you guessed it. The station went off the air again for half an hour!

She called back the next day and this time Pat told her she could talk as much as she wanted, but she said there was no need because the aliens had made their point.

A remarkable coincidence? A publicity stunt? Someone actually hacking the transmitter? The public never heard a reasonable explanation as to what really happened. What I do know is that the bizarre affair triggered my interest in "aliens," and much to my surprise, I found that the local library had quite a collection of books on the subject. I read about the Roswell incident, and about all sorts of UFO sightings. By this time, I had developed an interest in science and found the "proof" for alien visits less than compelling. Many of the accounts were fanciful, and it seemed to me that the writers were sometimes driven more by commercial appeal than by evidence. This led me to look at all news reports, especially in the scientific realm, with a skeptical eye, and I took to evaluating them in terms of adhering to the tenets of responsible journalism.

These days, maintaining a skeptical eye has turned out to be quite a challenge with the tsunami of information and misinformation we face on a daily basis. We are no longer talking about one TV channel but hundreds, with satellite radio we can access thousands of stations, and of course social media allows anyone to have a say on anything. As we witness on a regular basis, any twit can twitter. Then there are the millions and millions of posts ranging from those featuring sound science to ones that host the inane blather of scientifically confused bloggers to whom responsible journalism is a foreign concept.

I have been trying to battle such misrepresentations of science for a long time. It seems like only yesterday, but some thirty-eight years have rocketed by since I faced my first question from a listener on CJAD radio in Montreal. I was excited to be given a chance to enlighten the public about chemistry and figured I would be asked questions about how aspirin is produced, how baking soda works, how the birth control pill was developed, or the difference between natural and synthetic vitamin C. To me, this was chemistry. But the first question I had to deal with took a different tack.

"Is it safe to kiss your golf balls" was the confounding query. I didn't quite know what to make of this, but I soon learned that some golfers have the habit of giving their balls a friendly peck for good luck before whacking them. The caller's concern was that the balls might harbor some pesticide residue that could have an effect on his health. I offered the opinion that based on our knowledge of the toxicity of pesticides from animal studies, surveys of the health of golfers, determinations of the amount of pesticides that could be released from treated turf, and the brief exposure involved in romancing a golf ball, any significant effect was unlikely. Then I went on to qualify my remarks with the old adage that only death and taxes were certain.

Since those beginnings in 1980, I estimate I have dealt with over 10,000 questions on the air, ranging from ways to remove toilet bowl rust stains (phosphoric acid) to why opening a can of coffee beans triggers the smell of cooked turkey (no idea). But the largest category of questions has mirrored the golf ball query, focusing on risk. Over the years, the list of concerns has expanded way beyond pesticide residues on golf courses to fluorinated compounds, nanoparticles, sodium lauryl sulfate, caramel, flame retardants, acrylamide, formaldehyde, dioxane, dioxin, diesel fumes, benzene, trihalomethanes, mercury, parabens, antimony, gluten, cell phones, phthalates, bisphenol A, oxybenzone, basa fish, GMOs, lead, driveway sealants, hand dryers, fabric softeners, processed vegetable oils, carrageenan, azodicarbonamide, BHT, polydimethylsiloxane, perchlorates, isoflavones, and countless others.

My answers to questions about these issues haven't changed a whole lot; I emphasize the difference between hazard and risk. Hazard is an innate property, the propensity of a substance to cause harm, while risk is a measure of the potential that the substance actually does cause harm after taking into account type and extent of exposure and factoring in personal liabilities such as age, gender, and medical history. With time, I have become more and more aware of the challenges of coming to a conclusion about risk and how it basically comes down to making educated guesses.

These days, I'm often asked whether I think the public is better informed about science now than back when I started. More informed perhaps, but not necessarily better informed. When I first dipped a toe into the turbulent waters of science communication, there were no smartphones, there was no Google, no email, no Food Network, no Discovery Channel. Now we have all these, plus Dr. Oz, Joe Mercola, Gwyneth Paltrow, Jenny McCarthy, and Suzanne Somers dispensing their version of scientific wisdom. Electronic newsletters spew out tantalizing and seductive headlines ad nauseam: "The Antioxidant That's 6000X More Powerful Than Vitamin C," "Increase Youth Hormones by 682% — Grow Younger in 120 Minutes," "Alzheimer's Vanished in Days after Ohio Woman Ate This" (of course it will cost money to find out what "this" is).

Pseudoexperts like Vani Hari, who has anointed herself with the moniker "The Food Babe," take to the web to offer categorical advice about what food additives, cosmetic ingredients, household chemicals, genetically modified organisms, or pesticides to avoid based on anecdote, emotion, and a selective view of the scientific literature. Of course, the Internet has a positive side as well. Proper scientific literature is just a few keystrokes away, and there are outstanding websites such as Science-Based Medicine, NHS Choices, Sense about Science, and Quackwatch. Unfortunately these are not as popular as the absurd websites such as NaturalNews that serve up an assortment of ludicrous conspiracy theories and offer simple solutions to complex problems. It seems our efforts to improve the public's understanding of science are being trumped by the flood of Internet pseudoscience.

I'm painfully made aware of this every morning when I sit down to check my email. It takes a few minutes to delete the offers to repay me royally for helping a stranded tourist who has been robbed in some foreign country and the solicitations for friendship with Russian women. Next, I usually glance through the various "alternative health" newsletters to which I subscribe to see what "jaw-dropping cures" have been found by "cutting edge doctors" who traipse around the world searching out natural treatments that "extremely wealthy and powerful people do not want revealed."

Today, addressed as "Dear Unsuspecting Friend," which of course immediately arouses suspicion, I was informed about the work of one "brilliant, world-renowned MD" who has "solved the lethal riddle of the cause of high blood pressure, elevated cholesterol, problem blood sugar, bone loss, and sexual dysfunction." (The alternative world, it seems, is filled with "world-renowned MDs," and "maverick physicians" who "don't buckle under attacks mounted by the slash, burn, and poison-driven establishment.")

Dr. Fred Pescatore has "single-handedly" discovered the secret cure, yes cure, that "Big Pharma is desperate to keep tightly under wraps." It is a "precise combination of grape skins, lemon peels, and pine trees," but if we want to find out more about it we have to purchase his modestly titled book, *The Franklin Codex: A National Treasure Trove of Shockingly Simple Healing Miracles*.

In the book, we will also learn how we can avail ourselves of a substance that makes twenty million cancer cells go "poof" in a mouse and "actually works better than pharmaceutical cancer drugs." The substance, Alpha-G, has no "energy-zapping or nauseating side effects" but to get the real deal you need a reputable source. Wonder what that may be?

Alpha-G is an extract of the shiitake mushroom, also available as "active hexose correlated compound (AHCC)" that is not approved as a drug but can be sold as a dietary supplement. Some studies, all funded by the manufacturer, have shown activation of white blood cells that attack abnormal cells, but that is a long, long way from curing cancer.

As for diabetes, "you can forget about needles, sawdust, and grass clipping diets and potentially lethal blood sugar drugs." Dr. Fred's breakthrough allows you to eat "golden fried chicken and gooey chocolate brownies" because these can "actually help cleanse the body of diabetes." Of course that has to be in conjunction with his "Secret Super-Charger," which is a "delicious, natural, ultra-healthy plant extract with near magical health-promoting powers." After reading about how Dr. Fred's "simple, easy cures will free us from the drudgery of mainstream medicine," I thought it was time to move on from the drudgery of Dr. Fred.

Enter Dr. Al Sears, "antiaging pioneer, who at least twice a year leaves his clinic to travel the world, looking for medicinal herbs and plants to help his patients." Apparently a recent trip took him to Jamaica, where he came across some native fishermen who crumbled dried leaves and bark into water and then harvested the fish that floated to the surface. Strange that this herbal expert who "blows away conventional medical wisdom" had not heard of plants like the Jamaican dogwood, which puts fish to sleep. In any case, he couldn't wait to tell his "research team" about his discovery and get Jamaican dogwood into the hands of his patients suffering from sleep problems.

But more careful research by his "team" should have revealed that Jamaican dogwood contains rotenone, a compound that not only stuns fish, but can kill them. Rotenone has also been used as an insecticide, but it is being phased out because of toxicity, particularly because of a possible link with Parkinson's disease. There's probably not enough rotenone in Dr. Sears's recommended "all natural Native Rest" to cause harm, but nobody really knows because the contents of such supplements are not regulated in the same way as drugs, even though they claim pharmaceutical effects.

Next, I opened a colorful newsletter that promised a once-in-a-lifetime chance to save me and my loved ones from diabetes, heart disease, cancer, arthritis, and more. Dr. Jonathan Wright, one of "the founding fathers of natural medicine, the 'top of the mountain' expert — the one all the others look up to and learn from" would reveal "forbidden information that has been suppressed for decades." Wright, we are told, has pored over stacks of hushed-up studies, hundreds of "underground" medical texts (voracious reader this man is), and has carefully distilled the cutting edge discoveries that can reverse disease in the "mother lode of healing secrets," his "Treasury of Natural Cures."

The newsletter does offer some clues to whet our appetite. The solution to arthritis is cetyl myristoleate, a substance isolated in 1964 from Swiss albino mice that escape arthritis. Since then, a number of clinical trials in humans have concluded that it is safe enough, and it may provide relief to some. In spite of vast literature on cetyl myristoleate, Dr. Wright claims that we've never heard about it "because when a natural substance works too well it goes on the 'blacklist' and the only way to learn about it is from someone with inside knowledge."

This "brilliant mind" has also discovered that "you could rub breast cancer out of your body" with iodine, that an extract of the *Berberis aristata* plant can "eliminate high blood sugar, slash bad cholesterol, and send triglycerides plummeting." Why have we never heard of it? Because "Big Pharma stands to lose about $70 billion a year if we find out." Actually, my files contain lots of studies about this plant, and aside from some intriguing research with animals, there is no indication that "this single natural extract could replace blood sugar and cholesterol-lowering drugs."

Before I could get on to the rest of my emails, the computer pinged with yet another newsletter, this one about how "the leader of natural medicine's new wave" and "one of the most sought-after Doctors of Naturopathic Medicine in the World" found a cancer cure that left "oncologists stunned." "Dear Friend," the newsletter began, "could the Holy Bible contain the secret to eliminating the worst disease of mankind?" According to Dr. Mark Stengler, the secret is to be found in Matthew 4, apparently something to do with "man shall not live by bread alone." Stengler's conclusion is that the answer to wiping out any cancer in a month is a low carbohydrate diet. Yeah.

Next on the email list was a report about Dr. Oz's show the previous day. By comparison, that promised to be rational. Not so. Oz's "Two Day Holiday Detox" included a way to flush "fat-promoting toxins" from the body with cabbage. Nonsense, it seems, doesn't take a holiday.

INFOMERCIALS PROVIDE SLANTED SCIENCE

Sometimes when I have a touch of insomnia, I'll turn on the TV. Cruising through the channels I came across Larry King, sporting his trademark suspenders, interviewing a guest on a program called *Larry King Special Report*. I knew Larry had left CNN and was hosting a couple of interview shows on the RT (Russia Today) and Hulu channels, but I don't get these, so what was I watching? It soon became apparent that this was not a legitimate interview show but rather an "infomercial." King was shilling for a dietary supplement, Omega XL, going on about how it provides miraculous joint pain relief. His "guest" was Dr. Sharon McQuillan, waxing poetic about how she recommends Omega XL to all her patients to "help protect their hearts, preserve their heart and vascular health." Larry, who has a history of heart disease, asked McQuillan how Omega XL can reduce the risk of heart attacks. She answered that "thirty years of studies have shown the benefits of omega-3s." That is true, but totally misleading since none of those studies used this product.

Some studies have indeed shown that eating foods rich in omega-3 fatty acids may lower the risk of death from heart attack. And trials with supplements containing DHA and EPA, the two major omega-3 fats found in fish, have suggested a benefit for people who have previously suffered a heart attack. For example, in a placebo-controlled trial, patients taking an omega-3 supplement were 6 percent less likely to suffer a decline in heart function as determined by magnetic resonance imaging (MRI) than those taking a placebo. That's not a very big difference, and they were taking four grams a day! Omega XL contains 6.3 milligrams of EPA and 4.9 milligrams of DHA, roughly 1/400th the dose that showed a minor benefit in the study! In other words, there is no basis to suggest that the EPA and DHA in this supplement can help protect the heart.

Contrary to the image projected by the "interview," Omega XL is not a DHA/EPA supplement. It is an extract of the green-lipped mussel found in New Zealand and is a complex mix of many compounds. There is some evidence for an anti-inflammatory effect, and for possible benefit in treatment of arthritis and perhaps even asthma, but Dr. McQuillan's enthusiastic recommendation for protecting the heart is not supported by evidence. As is revealed in the credits after the show, McQuillan, who is a GP specializing in "integrative, regenerative, and aesthetic medicine," was paid for her appearance.

Of course, Larry King is not the only celebrity who has lent his or her name to promoting a product. In fact, back in 2000 when Larry was still on CNN, he featured Olympic champions Dorothy Hamill and Caitlyn Jenner (still Bruce back then) as guests. *Larry King Live* was certainly not an "infomercial"; it was one of the most respected and most watched interview shows on TV. Both guests were there to talk about the pain-killing drug Vioxx. "My doctor prescribed Vioxx for me, and it's as if I've been given a new life," Hamill told King. "It's just, it's been amazing. I feel twenty years younger." Jenner had won the decathlon at the 1976 Montreal Olympics but subsequently had knee surgeries and shoulder problems that were resolved with Vioxx. Both athletes were paid by Merck, the drug's manufacturer, something that was made clear on the program.

Neither Jenner nor Hamill could have known at the time that Merck was already investigating an apparent increase in heart attacks among people taking Vioxx. Four years later, the drug would be recalled for that very reason, precipitating some 35,000 lawsuits and payments of over $4 billion by Merck to the plaintiffs. Were there people who were prompted by Hamill and Jenner's appearance to ask their doctors to prescribe Vioxx? Undoubtedly. Did some suffer adverse consequences? Who knows? But a supposedly objective interview show is no place for paid celebrity product endorsers.

There is also the thorny question of television ads for prescription drugs featuring celebrities. Legendary golfer Arnold Palmer and basketball star Chris Bosh both had health problems that required treatment with an anticoagulant and in a TV ad they sang the praises of Xarelto (rivaroxaban), an effective oral blood thinner. Although this is a prescription drug, and the side effects and risks are outlined, the ad still indirectly suggests taking medical advice from a celebrity who has no relevant expertise. Kim Kardashian West, who is famous for being famous, promotes Diclegis, a morning sickness relief pill for pregnant women. While the drug is effective, Kim's lighthearted tone on Instagram, "OMG have you heard about this?" prompted the FDA to censure her for not including risk information or limitations for the use of the drug, something she subsequently corrected. Was there any damage done? Who knows?

While infomercials and ads are not exactly educational, TV is not a total write-off. Television can sometimes provide interesting and unexpected insights into science.

SEINFELD HEATS UP

I'm an unabashed fan of *Seinfeld* and watch reruns regularly even though I've probably seen each episode a dozen times. I especially like the segments that poke fun at some wacky real-life scenarios. Kramer burning himself when he attempts to sneak coffee into a theater by hiding it in his pants and then suing the coffee shop for selling coffee that is too hot is a great example. "You're gonna walk out of the courtroom a rich man," Kramer's lawyer tells him. Of course the suit doesn't pan out, which comes as no surprise since the episode is designed to mock frivolous lawsuits.

When this episode originally aired in 1995, there was much talk of "tort reform" with then Texas Governor George W. Bush championing the cause. "Tort" derives from the Latin word *tortus* meaning "wrong" and refers to laws that allow an injured person to obtain compensation from whoever is deemed to have caused the injury. The reforms that were sought were basically aimed at protecting businesses and doctors from frivolous lawsuits.

The *Seinfeld* episode was sparked by the real case of seventy-nine-year-old Albuquerque resident Stella Liebeck, who sued McDonald's after spilling hot coffee in her lap and was awarded $3 million by a jury. Ms. Liebeck's case became a poster child for frivolous lawsuits and was welcomed by comedians who saw it as an absurd abuse of the legal system, worthy of scathing witticisms. Politicians also weighed in, warning that such lawsuits would result in price hikes because companies would pass on their legal expenses to consumers. The media just loved the story of the little old lady who spilled coffee on herself and seized the opportunity to get rich by suing McDonald's. But there was one little problem. The media got it all wrong.

One of the most important lessons I have learned over many years of dealing with scientific issues is that they inevitably become more complicated when you start scratching at the surface to see what is underneath. *Liebeck v. McDonald's* is a classic case of an incredibly distorted story that should never have been the subject of ridicule. This was not a case of a greedy lady launching a frivolous lawsuit over some minor injury that was caused by her own carelessness. It was a case of an extremely severe injury that could have been avoided had McDonald's made appropriate adjustments after having received over 700 complaints about serving coffee that was too hot.

Contrary to many media reports, Ms. Liebeck was not attempting to drive while holding the coffee between her knees. The car, driven by her grandson, was stopped in the parking lot, and she was in the passenger seat. The coffee spilled as she tried to remove the lid of the cup while holding it between her legs causing terribly severe burns, well documented by disturbing photographs. She required surgery and skin grafts, with her medical bills quickly exceeding $10,000. Ms. Liebeck had no thoughts of suing, but she contacted McDonald's to inform the company that she had been told by her surgeon that any liquid in the range of 180 to 190°F will cause second- or third-degree burns if skin contact is more than a few seconds. She asked that the company check its equipment to ensure that the coffee was not being served at an excessively high temperature and asked for help with the payment of her medical bills. Only when McDonald's offered the paltry sum of $800 did the family sue.

A mediator recommended that the company settle for $300,000, but that was rejected and the case went to trial. The jury awarded Liebeck $160,000 for expenses plus $2.7 million in punitive damages with the intent of punishing McDonald's for not having addressed the hundreds of previous complaints it had received about excessively high coffee temperature. The judge reduced this amount to $480,000, bringing the total the victim received to $640,000, not the $3 million reported by the media. Ms. Liebeck never completely recovered her health.

Curiously, the case actually worked out pretty well for McDonald's. Mostly due to inaccurate reporting, the public sided with McDonald's, noting that it was a place where a nice hot cup of coffee could be had at a reasonable price. The verdict did lead to warnings on coffee cups about the contents being hot and to the design of cups less prone to spilling. Now I'll leave it for you to decide whether the lawsuit was frivolous.

And while contemplating that, how about a comparison with a suit that talented singer, actress, and producer Barbra Streisand filed in 2003 against the California Coastal Records Project? This project uses a photographic database of over 12,000 pictures of the California coast for a scientific study of changes in the coastline. Streisand claimed that a single photo that included her Malibu estate invaded her privacy, violated the "anti-paparazzi" statute, sought to profit from her name, and threatened her security.

The suit ended up in the Los Angeles Superior Court and was dismissed with the court finding that Streisand had abused the judicial process and ordering her to pay the defendant's legal fees. Before the lawsuit hit the media, six people had viewed the picture of Barbra's house on the website. After the story of the lawsuit made headlines, thousands of people flocked to the website. I bet the *Seinfeld* show writers could have come up with a witty episode about the "The Streisand Effect," a term now used for any enhanced, unintended publicity that is generated when an attempt is made to hide, remove, or censor a piece of information in the public domain. Alas, *Seinfeld* went off the air in 1998. Thank goodness for reruns.

LAUNDRY AND TV SLEUTHS

Columbo is another one of my favorites. We last met the disheveled police detective in 2003, but like Seinfeld, he is still alive in reruns. Lieutenant Columbo gave the appearance of being somewhat scatterbrained, but he was actually very clever, with great powers of observation as demonstrated in an episode entitled "Uneasy Lies the Crown."

"I've never been very good at chemistry," a dentist reveals to Lieutenant Columbo in that episode. And that remark would eventually do him in. On the other hand, Al Stewart, a real-life traveling salesman *was* good at chemistry, and that made him a fortune. What links the dentist and the salesman? Ferric ferrocyanide, better known as liquid bluing.

Stewart was a traveling salesman in the 1870s, selling groceries in southern Minnesota. He had heard that European women were adding some sort of blue coloring to the rinse water after they had laundered their white fabrics and figured selling this would be a good sideline. As fabrics age, their molecules undergo chemical changes so that they no longer reflect all the colors of the visible spectrum. They absorb some of the blue wavelengths with the result that the fabric starts to take on a yellowish hue. The addition of small amounts of blue coloring can compensate for the absorption of the blue wavelengths, and the fabric will again look white.

A little research by Stewart revealed that the European secret was Prussian blue, or "ferric ferrocyanide," a chemical accidentally discovered in the eighteenth century by Johann Dies
a Prussian color merchant who needed some potash
his dyes. He purchased this from Johann Konr
alchemist who had achieved some fame for r
venating" potion. But when, according
Diesbach mixed Dipple's potash with
sulfate), he got an unexpected bri
the potash had been contamina
Dipple used to make his tonic. It s
green vitriol provided the needed iron

Stewart was adept enough at chemistr

and began selling laundry bluing. Eventually it grew into a huge business, and Mrs. Stewart's Liquid Bluing was made available across Canada and the U.S. Today, it is still making our whites whiter, but Mrs. Stewart's bluing has other talents as well. It can be added to shampoos to make gray hair look brighter and to swimming pools to make the water look bluer. It can even save the lives of chickens. The temperature in poultry-rearing houses can shoot up in the summer, and at 82°F, about 6 percent of the chicks die from the heat. A common technique to prevent this is to whitewash the roofs of the chicken houses to reflect the sunlight. It turns out that the addition of a little bluing increases the efficiency of this process.

There's something else that bluing can do. Catch murderers. At least on TV. The dentist Lieutenant Columbo encounters in "Uneasy Lies the Crown" carries out his ingenious murder by placing digitalis in the crown he is using to cap the victim's tooth. A time-release gel secures the drug, preventing an instant reaction. Indeed, the plan works and the victim dies of digitalis overdose later that day. Digitalis overdose mimics the symptoms of a heart attack.

Chemistry plays an interesting role in the episode, extracting a confession from the murderer. Columbo is quite sure the dentist is the killer but can't think how he could have administered the digitalis until he overhears a waiter talking about treating his cold with a time-release medication. That implants an idea. Is it possible to use some sort of material in a crown that will allow a hidden drug to be released later?

The hallmark of each episode is Columbo's apparently friendly relationship with the suspect, often in the guise of asking for help to solve the crime as he attempts to elicit incriminating evidence. In this case he questions the dentist about time-release gels, but all he gets is a comment about never having been good at chemistry. That comes in handy when he plans to spring a trap. The idea for that trap germinates when Columbo notes some blue stains on his shirt left over from laundry bluing.

While playing the game of soliciting an expert opinion, Columbo invites his suspect to an autopsy of the victim's exhumed body, explaining that he believes the poison had been planted in the dental crown and saying he has a way of proving it. In the autopsy room, the detective reveals a chemistry set and describes that he has been able to determine that porcelain, as found in the crown, reacts with digitalis to produce a blue color. He demonstrates this by taking a crown and treating it with what he claims is digitalis. Indeed, it turns blue.

Of course the "digitalis" is actually ferric ferrocyanide, the blue pigment used in Mrs. Stewart's Bluing. At this point, Columbo makes his accusation and tells the suspect that he will now remove the crown from the victim and see if it has turned blue. That never comes to pass because the murderer sees he has been cornered and confesses his guilt. Good thing, because the crown would not have been blue since porcelain and digitalis don't actually react. Justice is served thanks to the criminal's ignorance of the chemistry of porcelain.

And just who was Mrs. Stewart? She was Al's wife, who helped him make the bluing agent. He thought that the picture of a housewife on the bottle would inspire confidence and increase sales. But the Mrs. wanted no part of the scheme, so Al plucked a picture of her mother from the mantle and used that. The kindly old lady who has been looking back at us from the bottles of Mrs. Stewart's Bluing for over a hundred years and has been helping with our laundry is not Mrs. Stewart, but her mother!

THE MYSTERIOUS ISLAND

"You are what you eat" is a time-honored truism. After all, food is the only raw material that enters our body, so we are literally made of what we eat. That of course includes our brain. But what do we fill that brain with? Here I would propose another maxim: "You are what you read." That comes to mind because I was recently asked how I originally got interested in what I do. That meant I first had to think about what it is that I do. Of course, I know what I spend my time on. I teach, I write, I blog, I answer email questions, I'm active on Facebook, radio, and TV, always with an emphasis on science. But what is at the heart of what I really try to do? Simply put, I think I try to demystify science and dispel myths by sticking to facts. My views and approach have evolved over the years, but there was a germinating factor.

I grew up in Hungary before the Internet, before computers. We didn't even have a telephone. I didn't know that television existed until I came to Canada. We did have radio, and I actually remember hearing the announcement of Stalin's death, and listening to the 1954 World Cup Final between Germany and Hungary. The Germans got lucky because Puskás was injured.

So what did I do in the evenings? I read books. And some I think played a significant role in formulating my future interests. I was absolutely captivated by the first novel I ever read, Jules Verne's *The Mysterious Island*. It told the story of how a group of northerners captured by the Confederates during the Civil War escaped by hijacking a balloon and became marooned on a deserted island somewhere in the South Pacific. Stranded there, they are forced to establish a colony by making use of their wits. One of the castaways, Cyrus Smith, is an engineer who turns out to be sort of a forerunner of MacGyver, the TV hero whose encyclopedic knowledge of science helped him solve problems by making use of whatever ordinary materials were available. Guided by Smith's profound practical knowledge of botany, geology, physics, and chemistry, the colonists fabricate cooking pots and bricks, manage to smelt iron, and even design a primitive telegraph system on the island.

Much to their surprise, as they run into problems, mysterious solutions appear. A box filled with weapons and tools inexplicably materializes, tablets of quinine magically turn up when malaria strikes, and a horde of invading pirates end up dead without any visible wounds. With no logical explanation apparent, it seems that some benevolent deity is looking out for the colonists' welfare. But in the end, the mystery is solved. The island turns out to be the hideout of Captain Nemo, a scientific genius, who lives in a grotto aboard his submarine, the *Nautilus*. It was he who had been the settlers' mysterious benefactor. All of the events that had been so puzzling now turn out to have a down-to-earth explanation. At that young age, I didn't understand all the scientific details described in the book, but a couple of points struck home. Scientific ingenuity can solve a lot of problems, and phenomena that at first seem paranormal can turn out to be quite mundane as facts come to light.

One of the problems the colonists faced was the need to find a source of water. With his knowledge of geology, the engineer locates an underground lake. Unfortunately, it is inaccessible. An explosive would be needed to blast apart the rocks blocking the path to the water. Smith has an idea. Make some nitroglycerin! And that, I think, was my first exposure to chemistry. Although I didn't follow Verne's description of the process of making nitroglycerin, I do know that subsequently my interest perked every time I ran across the term. When I heard that *The Wages of Fear*, an adventure film about the difficulty of transporting the compound, would be playing in our local cinema, I begged my parents to take me. (Yes, we did have movies.) Later, I would often write about nitroglycerin, an excellent example of how a chemical can be used either to the benefit or detriment of mankind.

Recently I reread *The Mysterious Island*. With my understanding of chemistry, I now marvel at Jules Verne's classic more than ever. His description of Smith's production of nitroglycerin is brilliant and scientifically plausible. The key chemicals needed are glycerin and nitric acid and Smith manages to make both.

The colonists' dog is attacked by a dugong, a manatee-like marine creature, and an underwater struggle ensues with the dog being saved by a mysterious hand (Captain Nemo's, as we later learn) that kills the dugong. The fatty animal is just what is needed to make glycerin. Any fat treated with soda (sodium carbonate) yields glycerin and soap, one of the oldest known chemical processes. But where to get sodium carbonate? It can be extracted from the ashes left when seaweed burns, which is just what Smith did. Then he needed nitric acid. That can be made by treating potassium nitrate, or saltpeter, with sulfuric acid. There was plenty of bird poop on the island, a good source of saltpeter, and fool's gold, or iron sulfide, was also abundant. Heating the sulfide converted it into iron sulfate, which when distilled yielded sulfuric acid. When it came to making a still, Smith's knowledge of pottery came in handy. The clever engineer then reacted the glycerin with nitric acid and produced the required nitroglycerin!

It turns out that the book that first stimulated my interest in science and sparked my passion for solving mysteries is more drenched in chemistry than I ever realized.

NUTTY SCARES ABOUT NUTELLA

There was panic in Italy. Some stores stopped selling Nutella, that creamy chocolate-hazelnut spread which is practically an Italian staple. It was removed from shelves after headlines claimed that "A Study Shows Nutella Can Cause Cancer." Actually, there was no such "study." The headlines were germinated by a report from the European Food Safety Authority (EFSA) calling attention to the presence of some by-products of palm oil processing that had raised concern because feeding them to rats caused a slight increase in the risk of cancer. Palm oil is the most widely consumed vegetable oil in the world, found in numerous foods, including ice cream, margarine, cookies, bread, instant noodles, baby formula, chocolate, and, yes, Nutella. But nowhere in the report was there any mention of Nutella. And neither did the report suggest that people should stop eating foods made with palm oil.

So why did the headlines feature Nutella? Because it is a very popular product, and any mention is guaranteed to capture readers' attention, which of course is the media's goal. Why is Nutella so popular? Because it tastes good! Let's face it though, a blend of sugar, palm oil, hazelnuts, cocoa solids, vanillin, whey, lecithin, and milk powder is hardly a health food, even though Nutella managed to associate itself with fitness through a sponsorship deal with the Italian national football team. In the U.S., the company actually ran into trouble with its claim that Nutella was "part of a nutritious breakfast," eventually agreeing to a $3 million settlement after being targeted in a class action suit for false advertising. But associating the product with cancer because of the possibility of the presence in trace amounts of some potential animal carcinogen is beyond the pale. First, a bit of history.

Cocoa was introduced to Europe by the Spaniards after landing in the New World. When the Duke of Savoy married Catherine, daughter of Philip II of Spain, in 1585, he fell in love with chocolate. As a result, Turin, at the time capital of the Duchy of Savoy, became a hotbed of chocolate consumption. It was in Turin that supposedly cocoa powder was first mixed with cocoa butter to make a chocolate bar. Fame of the solid chocolate spread, and Turin was soon supplying chocolate all over Europe. But a problem appeared when Napoleon spread his tentacles to Italy. The British blockaded ports they believed were used to supply Napoleon's armies, and this resulted in a severe shortage of cocoa powder imported from America. That's when Turin chocolatier Michele Prochet had the idea of extending chocolate with the hazelnuts that grew abundantly in the area to create blocks of "Gianduja." People took to spreading thin slices of this on bread.

After the Napoleonic wars, chocolate manufacture returned to Turin only to be impeded again by post–World War II rationing in Italy. That is when Michele Ferrero had an epiphany as he tinkered in his parents' small café with ways of making the small amount of chocolate available go further. Pietro, his father, had taken to selling Pasta Gianduja as a spreadable product, but customers complained that it didn't spread easily. So young Michele added palm oil to the product and came up with Nutella! By the time he died in 2015 at age eighty-nine, Michele had become a multibillionaire, listed as the twentieth richest man in the world. His company had added Kinder Eggs, Tic Tacs, and Ferrero Rocher chocolates to its offerings, but Nutella remains the flagship product, produced in eleven factories around the world and sold in 160 countries. In 2014, the

Italian Post Office issued a commemorative stamp honoring Nutella, and there is even a World Nutella Day, celebrated on February 5.

So it is understandable that when rocks are hurled at Nutella, people pay attention. These rocks are in the form of glycidyl fatty acid esters (GEs), 3-monochloropropanediol (3-MCPD) and 2-monochloropropanediol (2-MCPD), none of which are found in raw palm oil but are produced during processing. Palm oil is derived from the reddish pulp of the fruit of the oil palm but undergoes a fair amount of processing before it ends up in food. The raw oil contains residual chlorophyll and carotenoids that impart color, oxidation products that are smelly, and phosphatides that cause a gummy texture. "Degumming" involves adding citric or phosphoric acid to break down phosphatides and then heating with bleaching earth to bind the remnants. This process also removes colored and odiferous impurities. Bleaching earth is a type of clay that has been used since ancient times for decolorizing oils, but in the early 1900s, a German company discovered that first treating the clay with hydrochloric acid improved its performance. As it later turned out, it is the use of this acidified clay in conjunction with heat that results in the formation of the contaminants that were targeted in the EFSA document.

The formation of these contaminants has been widely studied, and it is well-known that they form only when the temperature is above 200°C. Manufacturers today, including Ferrero, keep the temperature below this and indeed there are no reports of any significant amount of GE, 3-MCPD, or 2-MCPD being detected in Nutella. On the other hand, researchers have discovered that cooking meat at a high temperature, especially charcoal broiling, results in fats forming glycidyl fatty acid esters (GEs) at higher levels than found in refined vegetable oils.

Maybe McDonald's was actually reducing risk with the introduction in Italy of Sweety, a meatless burger that is just a bun filled with Nutella. But now, with the nonsensical headlines about Nutella and cancer multiplying, it will be interesting to see if it stays on the McDonald's menu. I bet it would be a hit in America. Bottom line? There are reasons to limit Nutella consumption, but the supposed presence of carcinogens is not one of them.

DUBIOUS TIDINGS OF DOOM

"Half of All Children Will Be Autistic by 2025, Warns Research Scientist at MIT." That headline has scooted around the Internet since 2014, triggering both fear among the public and scathing attacks about irresponsible fearmongering by scientists. So, why are we destined for such a tragedy, and just who is this prophet of doom at MIT?

This story is really all about glyphosate, the most widely used herbicide in the world. Back in 1970, Monsanto, which was then a chemical company, was searching for novel water-softening substances. It was chemist John E. Franz's task to come up with some candidates. As is often the case, when new compounds are synthesized, companies run them through a battery of tests to see what other applications they may have. Two of the new compounds had weak herbicidal activity, and Franz was asked to synthesize some analogs that would have a more potent action. That research yielded "glycine phosphonate," a name that was quickly contracted to "glyphosate" and commercialized as "Roundup." The impact of glyphosate on agriculture was such that Franz received the prestigious Perkin Medal for outstanding work in applied chemistry in 1990.

Glyphosate proved to be very useful in ridding vineyards, olive groves, orchards, parks, and roadsides of weeds and aroused no public interest until Monsanto introduced seeds that had been genetically engineered to yield crops resistant to the herbicide. Now fields of genetically modified "Roundup ready" canola, soybeans, corn, and sugar beets could be sprayed with glyphosate to eliminate weeds without harming the crops. Historically, the introduction of any new technology, be it pasteurization, vaccination, microwave ovens, or cell phones, has raised concern, and so it was with genetically modified organisms (GMOs). There were allegations that the effects of GMOs on people's health had not been adequately tested and that we were all "guinea pigs." But scientific organizations and regulatory agencies around the world dismissed the concern that genetic modification alters the composition of the edible portions of these plants in any significant way.

Recently, worries about safety have expanded to include the supposed toxicity of glyphosate itself and its residues in food. That concern was spawned by the International Agency for Research on Cancer's (IARC) declaration that glyphosate is a probable human carcinogen. As was quickly pointed out by many scientists, this conclusion, while technically correct, was misleading because it did not take into account exposures that were relevant to the public. Independent agencies such as the National Academy of Sciences in the U.S., the United Nations Joint Meeting on Pesticide Residues, the Food and Agriculture Organization of the United Nations (FAO), the European Food Safety Authority (EFSA), and the Health Canada Pesticide Management and Regulatory Agency (PMRA) reexamined the glyphosate issue and concluded that there was no evidence of harm. This did not satisfy activists, who claim that some of the scientists involved in dismissing the risk had conflicts of interest and that the potential risk of ingredients other than glyphosate in Roundup was not taken into account.

They point a finger at chemicals such as polyethoxylated tallow amine, a surfactant that allows glyphosate to penetrate leaves more readily. This chemical, they say, has a genotoxic effect on cells. While some test tube experiments on cultured cells have shown some abnormalities, there is no evidence that such surfactants, which are widely used in many foods and toothpastes, have any effect on people. But the overall scientific conclusion that glyphosate preparations are not harmful in the amounts to which people are exposed, has not stopped activists from raising fear.

Enter Stephanie Seneff, PhD, who is a Senior Research Scientist at the MIT Computer Science and Artificial Intelligence Laboratory with no expertise in toxicology, agronomy, or epidemiology. For some reason she has become convinced that glyphosate is the devil incarnate and has published articles linking the chemical to gastrointestinal disorders, obesity, diabetes, heart disease, depression, Alzheimer's disease, and infertility. But don't look for her papers in mainstream journals. You'll find them in "play for pay" publications such as *Entropy*, an "open access" journal that will publish almost anything as long as the fees are paid.

Seneff's main thesis is that glyphosate disrupts gut bacteria and interferes with cytochrome enzymes, but she presents no relevant human evidence. The paper is peppered with phrases like "we believe" and "exogenous semiotic entropy," three words that have never occurred together anywhere except in this paper. There are also flagrant attempts to snow people with a mass of irrelevant data to make a case for glyphosate being the curse of our lives. Seneff's most spurious argument, repeated in her public presentations ad nauseam, is the correlation between increased use of glyphosate and increasing rates of autism and celiac disease. What we have here is the classic fallacy of confusing "association" with "cause and effect." Instead of glyphosate, one could just as well link an increase in these conditions, which is itself contentious, with an increase in coffee consumption, cell phone use, flat-screen TVs, Chinese imports, or sales of organic produce.

In any case, even if glyphosate is an evil chemical, it cannot do its mischief without exposure. So here is some data that is relevant. Based on extensive cellular, animal, and human epidemiological research, the acceptable daily intake (ADI) of glyphosate has been determined to be 2 milligrams per kilogram of body weight (0.5 milligrams in Europe). This means that a 50 kilogram person could take in 100 milligrams a day without any effect. Given the known relationship between intake and urinary excretion, at the above level of intake this would result in 15 milligrams being detected per liter of urine. Studies have shown that the actual secreted amount in the general population is 1 to 3 micrograms per liter! This corresponds to 1/5000th of the ADI, which actually has a 100-fold safety factor already built into it. Essentially then, our exposure to glyphosate as a residue in food is insignificant. So, if we are looking for the causes of our ailments, we need to look elsewhere. I'll also gladly wager Dr. Seneff that half of our children will not be autistic by 2025.

BLOWING IN THE WIND

David Copperfield performed many an illusion on his television specials with his hair blowing in the wind, tussled by an off-stage fan. I was reminded of that effect by an episode of *The Dr. Oz Show* in which the hot air so often generated by the host was amplified by a fan à la Copperfield. And Oz too was performing sort of an illusion if we go by the definition of the term as "something that deceives by a false perception or belief." In this case, Oz dumped a bunch of yellow feathers on a patch of synthetic turf adorned with some synthetic plants to demonstrate pesticide drift. The flurry of feathers was meant to illustrate how neighboring fields, as well as people who happen to be nearby, may be affected. A powerful visual skit to be sure, but a gross misrepresentation of the risks posed by pesticide drift.

The reason for the demo at this particular time was that, in Oz's words, "the Environmental Protection Agency is on the brink of approving a brand new toxic pesticide you don't know about." The reference was to Enlist Duo, a pesticide that at the time was already approved in Canada. It is actually a mixture of the weed killers glyphosate and 2,4-dichlorophenoxyacetic acid (2,4-D), designed to be used on corn and soy grown from seeds genetically engineered to resist these herbicides. Fields can then be sprayed to kill weeds without harming the crops.

The need for the new combination was generated by the development of resistance to glyphosate by weeds in fields planted with crops genetically modified to tolerate this herbicide. Such resistance has nothing to do with genetic modification; it is a consequence of biology. Some members of a target species will have a natural resistance to a pesticide and will go on to reproduce, yielding offspring that are also resistant. Eventually the whole population becomes resistant. This is the same problem we face with bacteria developing resistance to antibiotics.

Oz got one thing right. Pesticides are toxic. That's exactly why they are used. And that is why there is extensive research about their effects, and strict regulation about their application.

Remember that there are no "safe" or "dangerous" chemicals, just safe or dangerous ways to use them. But there's nothing new here. Both 2,4-D and glyphosate have been widely used for years, although not in this specific combination. What is new is the development of crops resistant to 2,4-D, which will allow for its use to kill weeds in corn and soy fields, something that was not possible before. This has raised alarm among those who maintain that 2,4-D is dangerous and that its increased use will affect human health. Dr. Oz is apparently of this belief, and as the feathers were flying around the stage, he chimed in with how "2,4-D is a chemical that was used in Agent Orange, which the government banned during the Vietnam War."

2,4-D was indeed one of the components in the notorious Agent Orange used to defoliate trees in Vietnam. Tragically, it was later found to be contaminated with 2,3,7,8-tetrachlorodibenzodioxin (TCDD), a highly toxic chemical linked to birth defects and cancer. This dioxin, however, has nothing to do with 2,4-D. It was inadvertently formed during the production of 2,4,5-trichlorophenoxyacetic acid (2,4,5-T), the other component in Agent Orange. That is why the production of 2,4,5-T, but not 2,4-D, was banned!

It is deceitful to imply that the new herbicide is dangerous because it contains the harmful compound that was used in Agent Orange. Not only does Enlist Duo not contain any TCDD, the form of 2,4-D it does contain is also different from what was used in Vietnam. Enlist Duo is formulated with 2,4-D choline, which is far less volatile than 2,4-D itself and has an even safer profile. While legitimate concerns can be raised about genetic modification, it is disingenuous to scare the public by linking the newly proposed herbicide to Agent Orange. It is also irresponsible to show videos of crops such as green peppers being sprayed, insinuating that Enlist Duo will be used on all sorts of crops whereas it would only be suitable for genetically engineered corn and soy.

Now on to the issue of pesticide drift, which can happen in two ways. Tiny droplets of the spray can be carried by air currents, and the chemicals can also evaporate and spread as a vapor after being deposited on a field in their liquid form. These are realistic concerns, especially given that some schools are located in the vicinity of agricultural fields. But these are just the sort of concerns that are taken into account when a pesticide is approved. For example, one well-designed study concluded that a person standing about 40 meters from a sprayer would be exposed to about 10 microliters of spray, of which 9 microliters are just water. Calculations show that the amount of 2,4-D in the 1 microliter is well within safety limits, and of course spraying isn't continuous. It is done a few times a year. Consider also that 2,4-D choline has far lower volatility and tendency to drift than 2,4-D itself, further improving its safety profile.

Another factor that is taken into account before a pesticide is registered for use is its mode of action. Glyphosate, for example, interferes with the synthesis of amino acids needed by plants to produce vital proteins. The pathway by which these amino acids are produced is not found in animals. Humans have no need for such biochemical synthesis because the amino acids we require are supplied by our diet. As far as 2,4-D goes, it mimics the action of a plant hormone and causes rapid growth of plants that cannot be sustained by available nutrients, causing the plant to wither and die. Such plant hormones have no human equivalent. This of course does not mean that these substances cannot cause harm by some other mechanism, but nevertheless the fact that their mode of action is through processes not present in humans is reassuring.

While no pesticide can be regarded as risk free, the portrayal of Enlist Duo by Dr. Oz amounts to unscientific fearmongering. His final comment that "this subjects our entire nation to one massive experiment, and I'm very concerned that we're at the beginning of a catastrophe that we don't have to subject ourselves to" totally ignores the massive number of experiments that have been carried out on pesticides before approval, based on a scientific rather than an emotional evaluation of the risk versus benefit ratio. True, when it comes to pesticides, there is no free lunch. But without the judicious use of such agrochemicals, producing that lunch for the close to ten billion people who by 2050 will be lining up for it becomes a challenge. What we need is rational discussion, not the spraying around of feathers and ill-informed rhetoric in a deception-laden stage act. If I want deception on the stage, I'll stick to watching David Copperfield.

FISH GENES AND TOMATOES

During a public lecture on genetic modification, I described an experiment that involved enriching soybeans with the amino acid methionine. Soybeans are widely used to raise animals but are low in this essential amino acid, often necessitating the use of methionine supplements. Brazil nuts produce a protein that is particularly rich in methionine, so the idea was to isolate and clone the gene that codes for the production of the methionine-rich protein and insert it into the genome of the soybean.

This raised an obvious concern. Although the modified soybeans were to be used mostly for poultry, the possibility that they could somehow end up in human food had to be considered. What if a person allergic to Brazil nuts happened to consume these soybeans, possibly triggering a life-threatening reaction? Testing of blood drawn from people allergic to Brazil nuts revealed that the antibodies they had produced in response to ingesting Brazil nut proteins also latched on to proteins in the engineered soybeans, indicating the potential for an allergic reaction. As a result, the research was abandoned and the modified soybeans were never produced.

The first comment after my talk picked up on the allergen issue. "If genetically modified foods were properly labeled, I could still eat tomatoes" was the angry remark. I was puzzled by this, but the gentleman went on to clarify. "I have a fish allergy," he said, "and I have no way of knowing which tomatoes have been modified with fish genes, so I just don't eat any tomato products." He need not have worried. There are no fish genes in tomatoes, and if there were, the tomatoes would have to be so labeled according to existing regulations. What we have here is fear generated by misinformation.

The Arctic flounder lives happily in the ice cold waters of the Arctic Ocean, its blood prevented from freezing by an "antifreeze protein." Since tomato growers live under the threat of a sudden freeze destroying their crop, researchers wondered about the possibility of inserting the flounder gene that codes for the antifreeze protein into the genome of the tomato. Preliminary experiments showed that in plants this protein was not effective in preventing ice crystal formation, and the project was dropped. But on the Internet, no story ever dies. The "fish genes in tomatoes" myth lives on, often illustrated with syringes plunged into tomatoes, or drawings of tomatoes shaped like fish. Had the technology proved promising, it would have required extensive testing of the specific fish protein used to determine if it was involved in producing an allergic reaction.

Such testing is not required when novel, conventionally-produced foods are introduced into the marketplace. Kiwifruit are an interesting example. Allergy to the fruit did not exist in North America until some thirty years ago, simply because kiwifruit was not eaten. With the expansion of global marketing kiwifruit are now found in every supermarket and, correspondingly, allergies have increased. Introducing a novel food, such as the kiwifruit, introduces hundreds of novel proteins, many with allergenic potential. On the other hand, genetic modification commonly introduces only one specific protein, meaning a reduced chance of an allergic reaction when compared to the introduction of new foods. This suggests that as far as allergies go, it is more important to focus on new foods, not on genetically modified ones. As people eat a wider variety of foods, they will develop a wider variety of allergies, but this problem doesn't get nearly as much attention as the potential reaction to a single protein in genetically modified food.

NEONICS AND BEES

Bees are critical to agriculture; there is no doubt about that. They fertilize various crops by spreading the pollen that they collect to meet their protein and fat needs. Recently there has been much concern about declining bee populations in some areas, and speculation has focused on insecticides known as neonicotinoids. Many media reports have tried and convicted the "neonics" and urged that they be banned. But as is so often the case, media reports only scratch the scientific surface, and deeper digging produces a different buzz. Neonics at a certain level of exposure can disorient or even kill bees — which comes as no surprise since they are insecticides, and bees are insects. The question is whether these chemicals should be banned outright or whether they can be used in a way that protects plants without harming bees.

Neonicotinoids, first introduced in 2004, are modeled on nicotine, the natural insecticide produced by the tobacco plant. One advantage of these chemicals is that instead of spraying, they can be applied to the seeds of crops such as corn, soybeans, and canola. They then end up distributed throughout the plant as it grows and are ready to dispatch any insect that dares to dine on the foliage. Bees don't do that, they go for the nectar in the flowers, which has only traces of neonics. Yet bee deaths have been linked to neonic-coated corn and soy seeds, mostly in Ontario. But curiously, not with canola seeds in western Canada, which are also treated with the same pesticides. So what is going on?

Mechanical planters use a jet of air to blow seeds into the soil. Commonly, talc or graphite is added as a lubricant to reduce friction between the seeds, but these can rub off and can carry insecticide-contaminated dust into the air, exposing flying insects such as bees to the neonics. The concern is that the tainted bees return to the hive where they can expose fellow bees to the neonics and wreak havoc. A novel polyethylene wax lubricant that can replace talc and graphite has shown a significant reduction in airborne insecticide during planting. There are also polymers being developed to help the insecticide stick to the seeds.

The planting of canola uses different technology and doesn't produce comparable amounts of dust. Some twenty million acres of canola are planted in Canada with neonicotinoid-treated seed and there has been no apparent impact on bee health. So it seems the problem may not be the neonics as much as the seeding methodology. Neonics are also commonly used on cut flowers and on plants purchased from nurseries but whether these affect pollinators is an open question.

In any case, the neonics are only part of the picture when it comes to bee health. There are mites, parasites, and viruses that can infect bees, and transporting hives, which is commonly done, also stresses them, as do harsh winters and long springs. Specifically, the Varroa mite can affect bee health significantly, and studies show that is especially the case when the bees' immune system is compromised by exposure to neonicotinoids.

So while the neonicotinoids may well be a factor in the decline of bee populations in some areas, they are not the only factor. Furthermore, loss of bee colonies has been observed in places where neonicotinoids are not used at all, and history records many cases of unusual deaths of honeybee colonies long before neonics were introduced.

Still, there are some troubling developments. A recent British study showed that bees are more attracted to a sugar solution laced with neonics than to one without, implying the bees may be getting some sort of a buzz from the chemicals and may be more likely to visit plants containing them and end up contaminating hives. And a study in Sweden showed a reduced density of wild bees, but not honeybees, in a field planted with neonic-coated seeds. Another large scale field trial examined canola grown in Germany, Hungary, and the U.K. either with or without neonicotinoid seed treatment and concluded that neonicotinoid exposure caused a reduced capacity of bee species to establish new populations in the year following exposure. A Canadian study determined realistic exposure of bees to neonicotinoids in corn-growing regions and used the data to set up experiments to study the effect of such exposures on honeybees. Worker mortality was increased and the colonies were more likely to lose queens.

Because of the cloud hanging over neonics, Europe and Ontario have decided to greatly restrict their use. It will take a while to see the effect, not only on the bees, but also on crop yields that have steadily increased since the introduction of the neonicotinoids. If yields are to be maintained, it may be back to the insecticidal sprays which come with problems of their own, not only for pollinators, but for people as well. Of course, in the Western world we can forego insecticides and just pay more for our locally grown food.

NATURAL FALLACIES

Drinking alkaline water can cure disease. Myth. Wrapping tarnished silver in aluminum foil and immersing it in hot alkaline water can remove the tarnish. Fact. Hot water with lemon juice is an effective "detox." Myth. Heavy metal poisoning can be treated with chelating agents such as ethylenediaminetetraacetic acid (ETDA). Fact. Autourine therapy can ward off disease. Myth. Organic agriculture allows the use of certain pesticides. Fact.

Separating myth from fact is the very essence of science and is the focus of many of my public presentations. It is common after a talk for someone to ask me what I think is the most prevalent myth I've had to confront over the years. Without doubt it is that natural substances have some sort of property that makes them superior to synthetic materials, with the corollary being that "natural" treatments as practiced by alternative practitioners such as naturopaths are preferable to the methods of "conventional" science.

"Natural" most definitely does not equate to safe. Natural coniine in hemlock put a quick end to the life of Socrates. In the eighteenth century, a local king in Java executed thirteen unfaithful wives by having them tied to posts and injected with the sap of the Upas tree through an incision on the breast. The latex from these trees contains antiarin, a potent cardiac glycoside. The Death Cap mushroom is well named, and tetrodotoxin in puffer fish, atropine in belladonna, or batrachotoxin in poison dart frogs can dispatch people pretty quickly. So can natural strychnine, botulin, or arsenic.

Aflatoxins in natural molds are potent carcinogens, and we are familiar with the effects of natural nicotine, morphine, and alcohol. Then of course there are the various pollens released by plants that annoy us with allergies and the myriad bacteria, viruses, and fungi that conspire to do us in with a host of dreadful diseases. And how about the mosquitoes that spread the natural malaria-causing parasite, the ticks that infect with Lyme disease, the snakes that inject a deadly venom, or the wasps that can double the size of your foot with their sting? The fact is that nature is not benign; even something as pleasant as sunshine can be deadly in the wrong dose. Natural radon gas is a carcinogen and poison ivy can create a great deal of misery. Visiting a urinal without washing your hands after handling hot peppers that harbor natural capsaicin will lead to a very memorable experience. Indeed, we spend a great deal of effort trying to outwit the natural onslaught with synthetic antihistamines, sunscreens, and chemotherapeutic agents. But some promoters of "natural" therapies also spend a great deal of effort trying to outwit us with pseudoscientific mumbo jumbo capitalizing on the "natural is better" myth.

Take for example the cleverly named dietary supplement 112 Degrees, promoted with the slogan "A new angle on sexual health." The geometric reference refers to the angle aspired to by men who suffer from erectile dysfunction. 112

Degrees claims to be a proprietary blend of "all-natural ingredients" that enhance male sexual vitality. While the advertising sounds pretty seductive, it is soft on hard facts. The inventor is Dr. Laux, who turns out to be a naturopath, not exactly the pedigree one looks for in a drug developer. He is presented as some sort of globetrotting knight in constant search of the best and safest "all-natural" treatments. Yup. How likely is it that someone with a smattering of scientific education is going to find an effective product that has eluded the giant pharmaceutical companies staffed by experts who scour the natural world for active ingredients?

The natural health industry commonly promotes the notion that pharmaceutical companies are not interested in natural products because they cannot be patented. This is not so. The use of a specific natural preparation can be patented just like a synthetic drug. Of course what really matters is not whether or not some substance is patented or whether it is natural or synthetic, but whether there is evidence to back the claims. 112 Degrees claims to be supported by numerous scientific studies. Yes, there are some studies, but they don't actually support the claim of enhanced male vitality. The studies show that the product is not carcinogenic, that it has some antioxidant potential, and that it has some ability to inhibit an enzyme that interferes with smooth muscle function. All good, but is there even one study to show that 112 Degrees can help men with erectile dysfunction? None that I can find.

The advertising refers to studies about some of the ingredients. *Butea superba* root for example. We are told that it was revered by royalty in the ancient kingdom of Siam for its power as an aphrodisiac. That is about as convincing as the story of ancient Assyrian men dusting their genitals with powdered natural magnetic stones and having their ladies follow suit by sprinkling natural iron filings across their own genitals for some literal attraction.

Then there is the claim that *Tribulus terrestris*, another herbal component, combats fatigue and low libido. No mention is made about how much is contained in 112 Degrees but we are reassured that Ayurvedic and early Greek healers used *Tribulus terrestris* as a sexual rejuvenator. One study, never duplicated, showed greater mounting behavior in mice, but there are no human studies that have shown any sort of effect on sexual performance or libido. There has been at least one report of breast growth in a man who took *Tribulus* as a weight training aid, for which it is in any case ineffective. In sheep, *Tribulus* has been noted to cause Parkinson's-like effects. Of course none of this is noted in the 112 Degrees documentation. So I think a large degree of skepticism, more than 112 degrees, is to be exercised when looking at the overexuberant and naïve promotion on behalf of this product by people who are trying to cash in on the unfounded "natural is better" notion.

NATURAL CURES

"I know you don't believe in natural medicine, but I'd rather take the medicines that nature provides than some synthetic drug full of side effects." So began an email, which then went on to list a variety of botanical supplements that supposedly help with every known ailment and ended with a tirade about the evils of "Big Pharma." I'm not exactly sure what was meant by the "natural medicine" that I don't believe in, but whether natural or not, medicine is not a matter of belief. It is a matter of evidence. It is fair to add, though, that a lack of evidence is not evidence of the lack of an effect. It may be just a lack of proper investigation.

There is no question that nature provides us with a bounty of substances with therapeutic potential. Indeed, more than half of the drugs used to fight cancer are derived from nature. Take vincristine for example, a drug used to treat some lymphomas and leukemias. Back in the 1950s, the Eli Lilly pharmaceutical company was following up on a folkloric treatment to control blood sugar with extracts from the Madagascar periwinkle. That turned out to be a dead end, but an extract of the plant did reduce activity in the bone marrow, signaling a possible treatment for leukemia. Like any botanical, Madagascar periwinkle contains hundreds of compounds and much research was needed before vincristine was isolated and shown to have anticancer activity. Suffice it to say that cancer patients are not told to dine on periwinkle or to swallow some extract of the plant. They are given an intravenous solution containing a carefully determined dose of the isolated and purified compound.

Another interesting example is eribulin, a drug used to treat complex cases of metastatic breast cancer. In 1986, Japanese researchers looking for biologically active substances isolated a compound from a sponge in the *Halichondriidae* family that they named halichondrin B. This had anticancer activity but was difficult to isolate. By 1992, Harvard researchers had managed to synthesize the compound in the laboratory allowing for more extensive investigation. It turned out that one specific part of the molecule was responsible for the relevant biological activity, leading to the synthesis of a molecule with a simpler molecular structure that incorporated the active fragment. This was named eribulin and introduced into the marketplace as Halaven.

With mounting concern about diseases spread by viruses, especially Ebola, researchers are exploring all possible sources of antiviral compounds. Perhaps a sort of "viral penicillin" will be found in something like honeysuckle tea, which according to traditional Chinese medicine is a treatment for the flu. There is a recent study in this vein that has generated a great deal of interest. The research showed that mice were protected from being infected by the H1N1 virus after drinking a decoction made by boiling mashed honeysuckle leaves in water. The researchers went on to isolate a compound, a microRNA, that they showed was capable of repressing the influenza virus by silencing two genes the virus needs to multiply. An interesting finding, especially since the same compound also targets the Ebola virus, at least in the test tube.

Then there is hibiscus tea. In a placebo-controlled trial, a cup taken with each meal reduced blood pressure from 129 to 122 mm Hg. And how about potato juice? After hearing a folktale about drinking it to treat stomach problems, University of Manchester researchers discovered that the juice was effective against *Helicobacter pylori*, the bacterium responsible for the majority of ulcers. Efforts are underway to identify and possibly isolate the active ingredients.

As is evident, nature, albeit usually with significant help from chemists, is capable of furnishing us with useful medications. But that doesn't mean we should swallow all claims about natural medicines hook, line, and sinker. Like the one about Blood of the Mountain, or *shilajit* in Sanskrit, a sort of goo oozing out of Himalayan rocks, supposedly the remnants of some ancient marine organisms. According to its exuberant marketer, it "reverses aging as it revitalizes sexual energy." What is the evidence? Older Himalayan monkeys are said to drink shilajit which makes them "act like energetic youngsters able to give their partners heart-pounding, thigh-quaking satisfaction." Sounds like monkey business to me.

LEG CRAMP RELIEF. REALLY?

Ayyayyayyayyah! That's perhaps the best way to describe a nighttime leg cramp. You wake up with an excruciating pain somewhere along your leg and all you can think of is getting relief. You massage, you pull, you push, you hop on one leg. Sometimes that helps, sometimes not. Then as suddenly as it came, the pain resolves. Big sigh of relief! But you never want to go through that experience again. So in the morning, you scurry to your computer to find out what Dr. Google has to say about this nightmare.

You quickly learn that you are not alone, especially if you are over sixty. Roughly one in three in that age group get night cramps somewhat regularly. Most of the time these are "idiopathic," meaning there is no known cause, although in some cases the cramps can be due to dehydration, imbalances in electrolyte levels, an underactive thyroid gland, peripheral vascular disease, or an adverse reaction to drugs such as diuretics, statins, calcium channel blockers, lithium, or the Alzheimer's medication donepezil.

As far as treatments go, there is scant evidence-based information. Walking on heels for a few minutes, or straightening the leg and pointing the toes towards the shin is said to help. Various exercises are offered to prevent cramps from occurring, the most common one being stretching the leg muscles by standing about a yard from a wall and leaning forward with arms outstretched to touch the wall while keeping the soles of the feet on the floor. Maintain this position for about five seconds and repeat as many times as possible over a five-minute period.

Quinine has been used in cases of very frequent recurring cramps but is rarely used today because of a host of possible side effects. Some people resort to tonic water because of its quinine content, but the amount is way too small to have any effect. There is some evidence that supplements containing the electrolytes sodium, potassium, calcium, and particularly magnesium, can help, especially in a chewable form for quick absorption. SaltStick Fastchews tablets, actually designed for use by high-performance athletes, have been praised by some nighttime leg cramp sufferers as a means of affording relief. That of course is what we call anecdotal evidence. It doesn't mean that it is untrustworthy, just that it hasn't been confirmed by randomized, double-blind trials, the gold standard of science.

Where science leaves a void, unconventional therapies rush to fill it. All you need to stop the cramps is an old Amish formula, one website claims. What is it? "A carefully balanced mixture of certified organic unfiltered raw apple cider vinegar, juice from the ginger plant, and just the right amount of all-natural garlic juice." Uh-huh. But without a doubt, the most unusual therapy promoted on the web is sleeping with a bar of soap. Just take a bar — some say it has to be Ivory, others maintain that any soap will do except for Dove or Dial — and place it on the mattress under the sheet. Pleasant dreams!

This is the kind of daftness we are tempted to immediately dismiss because it seems to make no sense at all. How could sleeping with a bar of soap have an effect on leg cramps? But the web serves up testimonials galore from desperate people who say that despite thinking this to be a ridiculous notion, they decided to try it anyway and found that it worked! Sounds like our good old friend, the placebo response. Might it not be the case that some people who have struggled long and hard with such cramps want so much to believe that something will help that they will respond to the presence of the soap? Some argue that soap fragrance may have some sort of relaxing effect, but this seems unlikely given that people swear by different kinds of soap and claim that it doesn't even matter if the bar is wrapped or not. Of course, if the sufferer feels better, it doesn't much matter why. So, I suppose there is no harm in telling someone that "some people believe that sleeping with a bar of soap helps" and suggest they give it a shot. That little white lie doesn't break the number one rule of medicine: "first of all, do no harm." It is hard to imagine how a bar of soap might do harm, although I suppose there is a chance it can be knocked to the floor by a leg cramp attack and someone could slip on it.

Escaping cramps by bedding down with soap is a slippery claim. So is the supposed relief afforded by Hyland's Leg Cramps PM, touted in its advertising as "the number one pharmacist-recommended brand for Leg Cramp Relief." When taken at the first twinge of a cramp, it is said to deliver quick help. This is a puzzler, because the product turns out to be a homeopathic remedy, which means that the "active ingredients" are present in an extremely diluted form. Most homeopathic remedies are so dilute that they contain nothing but water. This one isn't diluted quite to that extent; it actually contains traces of ten ingredients, none of which are known to trigger cramps in high doses. That is inconsistent with the basic theory of homeopathy, "like cures like." This pseudoscience is based on the fanciful notion that a substance that causes symptoms in a healthy person will alleviate those symptoms in a sick person when given in an extremely diluted form. Needless to say, there are no studies attesting to the efficacy of Hyland's Leg Cramps, so it is hard to understand how any pharmacist could recommend it. Just thinking of it gives me a mental cramp. But I suppose the stuff works as well as a bar of soap under the sheet.

THE POWER OF THE MIND

It's a fascinating story. And it may even be true. When a battlefield clinic ran out of morphine during World War II, a desperate nurse filled a syringe with saline solution and told a severely wounded soldier that he was getting a shot of a potent painkiller. His agony was almost immediately relieved! The amazing effect was duly noted by Dr. Henry Beecher, an anesthetist who was tending to the troops. He began to wonder to what extent the response to a medication was due to the power of belief rather than to any active ingredient.

After the war, Beecher returned to his post at Harvard and voiced criticism of the way drugs were being tested at the time. Basically, subjects were given a drug in various doses to determine its effectiveness and side effects. Beecher suggested that the only way to determine if the effects were due to the supposed active ingredient was to study a parallel control group given a fake pill. To make such a study truly objective, he proposed that neither the subjects nor the experimenters should know who was taking what. Furthermore, he suggested that the subjects should be randomly distributed between the experimental and control groups to ensure that the groups were as identical as possible. Beecher laid out his ideas in 1955 in a paper published in the *Journal of the American Medical Association* entitled "The Powerful Placebo," the term deriving from the Latin, meaning "I will please," which is exactly what placebos do. The article was instrumental in the U.S. Congress amending the Food, Drug, and Cosmetic Act, requiring that henceforth drugs be tested in double-blind, placebo-controlled randomized trials (RCTs).

Much has been learned about the placebo response since Beecher began his crusade. For example, the response is sensitive to cultural differences. In Germany, where people tend to worry more about low blood pressure than high, response to placebos to control hypertension is poor. On the other hand, placebo response to ulcer medications is strong because ulcer is a commonly diagnosed and treated condition in Germany.

Anticipation of an effect is also a factor, as is illustrated by Italian researcher Fabrizio Benedetti's classic 2003 study. Using a tourniquet, he induced severe arm pain in volunteers who were then told that they would be given either a powerful pain-killer or a drug that increases pain perception. In fact, all were given just a saline injection. Anticipating more pain led to more pain, anticipating relief delivered relief. Obviously, being told what to expect by a physician has an effect on what a patient experiences. This is one reason why alternative therapies with no scientific basis have success. The therapists have great faith in the treatments they offer, and that enthusiasm convinces patients of the effectiveness of the treatment.

One of the surprising aspects of placebo-controlled trials is that over the years, the response in the placebo group has become stronger and stronger, at least in North America, as recently pointed out by McGill researcher Jeff Mogil. This may be due to the widespread advertising of drugs here, creating an impression that virtually any ailment is treatable by pharmaceutical intervention. Although in a placebo-controlled trial the subjects do not know what they are taking, they do know that they may be getting a pill with an active ingredient. And because they have been bludgeoned by advertising touting the effect of pills, they are likely to evaluate placebos higher because of the belief that they may not be taking a placebo.

The placebo also has a wicked relative, termed the "nocebo," meaning that an expectation of side effects can increase the chance that these will manifest. In an Italian study, subjects with or without lactose intolerance were given pills they were told contained lactose, but in fact did not. Forty-four percent of the truly lactose intolerant subjects got diarrhea and stomach cramps. But so did 26 percent of the subjects who had never been diagnosed with lactose intolerance. They likely thought they were being tested for signs of the disease and that was enough to cause symptoms. In another study, half the men treated with finasteride for prostate enlargement were told that erectile dysfunction was a possible side effect, while nothing about such dysfunction was mentioned to the other half. Fifteen percent of the men who were not informed about the possibility developed erectile problems, whereas 44 percent of the men who had been advised of the risk experienced the problem. The mind is a powerful organ.

CONJURING UP REMEDIES

I collect magic memorabilia. Not only do I collect items, I often use them to make a scientific point. One of my prized possessions is a flowering bouquet I purchased at an auction that supposedly belonged to the legendary Harry Blackstone Sr., who featured it in the classic opening of the second act of his show, the Enchanted Garden. I can't be sure that it is authentic (although I paid for it as if it were) since no video of the famous performance exists. But the trick certainly does "perform." A green, flowerless bouquet is held in the hand, and then, magically, flowers slowly emerge and bloom. Ooohs and ahhhs all around!

I use the effect to talk about the real magical properties of plants. They're like chemical factories, taking in carbon dioxide, water, and various nutrients from the soil for conversion into fats, proteins, carbohydrates, and myriad other compounds. Why plants churn out hundreds and hundreds of compounds of varying molecular structure isn't clear, although many of them have been found to act as natural insecticides and fungicides. Indeed, in our food supply, which ultimately derives from plants, we ingest far more natural pesticides than the synthetic ones applied by farmers to increase yields. Among the numerous compounds produced by plants are many that have biological activity in humans. They can be poisons or they can be drugs, depending on the mode and extent of exposure.

Morphine, derived from the opium poppy, can be a wonderful painkiller, or it can kill by stifling respiration. Vincristine, from the Madagascar periwinkle, and paclitaxel, from the Pacific yew tree, are effective in the treatment of cancer. But of course patients are not asked to graze in a field of periwinkle or to chew on the bark of the yew tree. The active compounds are isolated, purified, tested, and standardized for medical use. Such plant-derived drugs have risen out of the ancient practice of herbalism, a fact vigorously pointed out by current marketers of "natural" products.

However, evidence that the complex mixtures sold as "natural health products" are effective is very thin. Take for example oil of oregano, a popular item in health food stores that is supposed to be good for just about everything. Take your pick of what it cures: sore throat, lice, colds, acne, infections, parasites, yeasts, diabetes, or allergies. On an episode of *The Dr. Oz Show*, the good doctor and his wife regaled us with the wonders of this herbal product. Lisa Oz has no scientific background but is a "reiki master," which apparently qualifies her to be an expert on herbal remedies. The couple went on about how carvacrol, the "super ingredient" in oil of oregano, destroys nasty bacteria and boosts the immune system. There was even a neat demo, in which a vile looking model of a bacterium was encased in what looked like a glass bubble. Dr. Oz attacked the bubble, which played the role of the bacteria's protective layer, with a kitchen knife.

The attack wasn't exactly a challenge to Norman Bates's efforts in *Psycho*, and was not successful. Then Lisa stepped in with a kettle of hot water, which played the role of carvacrol, and poured it over the bubble. It immediately cracked, and her knife-wielding hubby now easily burst through and punctured the bacterium, deflating it like a balloon. A really neat demo! I think they must have cooled the glass first to make it crack so easily. They get points for that one. Of course the point is way overhyped. There is some cursory laboratory evidence that bacteria perish when bathed in oil of oregano. But they also perish if bathed in a salt solution, alcohol, lemon juice, or a variety of soft drinks. It isn't hard to kill bacteria in a petri dish. But the body is not a large petri dish.

There is no evidence that a dose of oil of oregano is absorbed into the bloodstream to an extent where it may have an antibacterial effect, neither is there any evidence that it can treat any other condition. You would think that the promoters of the oil would be keen to organize randomized controlled trials to prove their point. But they are not. Why should they? People are buying the stuff anyway based on romanticized seductive anecdotes and a negative study would only serve to drown sales. What about the "immune-boosting" claim? Here, the evidence comes from nursing pigs. If they are given oil of oregano, they produce somewhat more white blood cells in their milk. Hardly something to oink home about.

What we have here are a few studies that suggest an effect in the lab or in animals that are then wildly exaggerated by marketers. It is often the case that the more claims are made on behalf of a product, the less likely it is that any of them are legitimate. Lisa and Mehmet Oz also promoted oil of oregano as a topping for toothpaste in order to kill germs in the mouth and actually had audience members brush their teeth with the concoction. There were contorted faces all around. Obviously, the wonders of oregano oil left a bad taste in the subject's mouth. Mine too. And I didn't even have any.

Researchers are of course interested in identifying compounds with possible physiological activity, subjecting them to laboratory tests, followed by animal experiments, and then possibly human trials. The desperately ill, on the other hand, are willing to ply themselves with herbal concoctions of unknown composition and efficacy, based essentially on faith in the provider of such remedies. Unfortunately, the hope, and even initial claims of success, quickly fade as magic is replaced by reality. Sort of like my bouquet.

The blossoms turn into flowers and then fall off. But then comes what in magic we refer to as the "kicker." Once more, the bouquet blooms, producing a fresh batch of flowers to replace the fallen ones! Just like in the world of plant medicines. As we lose faith in some of the failed remedies, invariably new hope emerges as some scientist discovers a compound that in preliminary studies lowers blood pressure, treats diabetes, or destroys cancer cells.

CANCER AND CARNY TRICKS

"Up your nose with a surgical clamp!" That is João Teixeira's prescription for treating breast cancer. I first came across this Brazilian "healer," known as John of God, in 2005 when he was featured on the ABC television program *Primetime Live*. John, who has all of two years of schooling, claims he is only an instrument in God's divine hands and that during a healing session his body is taken over by the spirits of long-dead physicians who guide his actions. Judging by the instructions they provide, it seems these physicians missed quite a few classes in med school. John, however, does not solely rely on departed physicians for advice; King Solomon can also be called upon when needed. The spiritual connections also allow John to diagnose a patient with just a glance.

Once the diagnosis has been made, the healing procedure begins. It may be "visible" or "invisible" spiritual surgery. If the patient chooses invisible, they are directed to a room to meditate while the spirits do their work. "Visible surgery" can involve sticking a surgical clamp up the patient's nose. It looks very impressive, but is nothing but an old carny trick, usually performed with a long nail and a hammer. Any anatomical text will reveal that there is a roughly four-inch-long passage up through the nasal cavity that is quite ready to accommodate a foreign object without any harm.

I recently saw an entertaining performance of this effect in front of the Ripley's Believe It or Not! museum in New York, of course without any implication of therapeutic value. The lady standing beside me gasped and exclaimed, "I don't believe it," despite just having examined the nail and having witnessed the show in front of her nose. It is easy to see how desperate people can be led by the nose to believe that some sort of supernatural power must be involved, and that someone capable of carrying out such a feat can perform other miracles as well.

John maintains that the success of his treatment hinges on the patient abstaining from drinking alcohol, eating pork, and having sex for forty days after treatment. That can provide for a convenient "out" in case no miracle occurs. Patients can be healed even if they are unable to travel to Brazil. All that is needed is a surrogate willing to undergo the spiritual surgery. No evidence for this remote healing is provided.

The forceps up the nose is not the only trick up John's sleeve. To treat nervous conditions, he appears to scrape the patient's eyeball with a knife while other problems are doctored with small random cuts on the body. As the *Primetime* cameras recorded, none of the patients showed any sign of distress after these rather invasive procedures. Quite the opposite. They believed they had been helped. Belief is a powerful tool indeed! There is a long history of television faith healers having the infirm throw away crutches and walk away unaided. Of course, no cameras are present when they crumple to the floor backstage. An adrenalin rush stimulated by faith can produce amazing effects.

In an attempt to provide a critical view of John's antics, the producers invited two experts, cardiac surgeon Dr. Mehmet Oz and James Randi, the world's leading investigator of "paranormal" phenomena. Oz was likely chosen because he was a proponent of various "alternative" therapies such as therapeutic touch and reflexology and would likely be somewhat sympathetic to faith healing and perhaps add an air of legitimacy. Randi was invited as the token skeptic.

Dr. Oz appeared repeatedly in the hour-long show, basically echoing the refrain that science doesn't have all the answers and that other forms of healing need consideration. Science of course doesn't claim to have all the answers, but it does look for evidence before jumping on a bandwagon. Randi, who could have provided evidence for methods of trickery and for psychological manipulation, was given a total of nineteen seconds on the show after being interviewed for hours. Why? Because the possibility that cancer can be healed by penetrating the nose with surgical forceps by a healer chosen by God makes for better television than declaring him to be a self-delusional simpleton or a calculating fraud artist.

In any case, it is a fact that people spend thousands of dollars to travel to Brazil to be poked, prodded, and scraped by John. Why? Because they are desperate and desperate people do desperate things. And many will provide alluring accounts of benefits. As Benjamin Franklin said, "There are no greater liars than quacks — except for their patients." Nobody wants to admit that they were swindled by some peasant who put tweezers up their nose. It is more comforting to believe that they were helped.

But what about the ones who give up conventional care to go this route because they believe it to be more effective? Like South African singer Lisa Melman, who refused breast cancer surgery to be treated by John in 2005 and appeared *The Oprah Winfrey Show* singing his praises. Unfortunately, in 2012 she stopped singing forever, succumbing to the disease of which she was supposedly cured.

In an ironic twist, in 2015 John of God complained of a pain in the stomach to his cardiologist. Yes, the medium who claims wondrous healing powers has a cardiologist, who without fanfare years earlier had implanted three stents in John's narrowed arteries. Now he sent his patient for an endoscopy that revealed a tumor. A ten-hour surgery, not the spiritual variety, was followed by extensive chemotherapy. A year later John appears to be well, cured not by mumbo jumbo, but by modern surgery and drugs. He had no problem affording the treatment; John is wealthy from donations and sales of blessed water and magic triangles.

When asked why he did not heal himself the way he is able to heal others, he replied with the stunning rhetorical question, "What barber cuts his own hair?"

NO MAGIC IN QUACK CANCER TREATMENTS

Magic is the science of fooling people for purposes of entertainment, and magicians, be they professionals or amateurs like me, take a great interest in the various methods that can be used to bamboozle people. Magicians, though, are honest charlatans and tend to get quite upset with dishonest charlatans who dupe people for purposes other than amusement. The most troubling forms of deception are the ones that deal with matters of health, particularly when it comes to preying on people's desperation. Cancer is a dreaded disease, and there are plenty of charlatans ready to take advantage of its victims. Over the years I have often tried to alert the public to the various ruses with which these unsavory characters ply their trade. But, as is often the case, personal involvement takes investigation to a whole other level.

When my wife was diagnosed with glioblastoma multiforme (GBM), a terrible type of brain tumor, I did what most people do. I frantically searched the scientific literature for options and quickly discovered that there was little room for optimism. Inevitably, Googling brings up a host of "miracle cures," ranging from herbal remedies and electronic gizmos to coffee enemas. "Cutting edge," "breathtaking," "bombshell report," and "more powerful than any drug Western Medicine can offer" are phrases often encountered. There are stories of "stunned doctors who watched tumors disappear in just two weeks," and accounts of patients who failed to improve with "dangerous chemo and agonizing surgery" but experienced a miraculous recovery after opting for a "little-known natural serum in a tiny vial that has the power to crush the billion-dollar chemo and radiation industry." To find out what it is, you are urged to watch a video, and to do so quickly, "before the government, conspiring with Big Pharma, will force its removal." But after you've invested close to an hour, you learn that you have to purchase a book or some newsletter to have the secret revealed.

Another report describes a man whose "body remained riddled with tumors after eight brutal months of chemotherapy and had already bought a grave before every single tumor in his body was obliterated." It costs to find out how. Then there are maverick physicians who claim to answer to the Hippocratic Oath, not drug companies, and want to "blow the lid off" Big Pharma's attempts to suppress a treatment "proven to be more effective than nineteen of their best selling drugs — but without side effects." There are numerous such websites featuring various "censored" cures, all claiming to have evidence that is being blocked from publication by drug companies trying to protect their turf. Right.

I received numerous emails from well-meaning people about treatments to explore, ranging from hemp oil and alkaline water to the Amazing Amezcua Bio Disc that promises to cleanse chakras. One, Light Induced Enhanced Selective Hyperthermia, seemed interesting enough to look into. What I found was not pretty.

Light Induced Enhanced Selective Hyperthermia was actually a scheme cooked up by Antonella Carpenter, an Oklahoma septuagenarian "alternative practitioner" who is not a physician but who has some training in physics. She claimed to cure cancer by injecting a tumor with a saline solution of food coloring and walnut hull extract followed by heating the area with a laser. The treatment, she maintained, was 100 percent effective with no side effects. Of course, any claim of 100 percent efficacy is a hallmark of quackery since no drug of any kind works in such a foolproof fashion. Even worse, Carpenter urged patients to stay away from oncologists and sometimes told them their cancer had been "killed," which was not the case.

As often happens, quacks unearth some legitimate process and then twist it out of proportion to hatch a money-making scheme. In this case, the legitimate process is photodynamic therapy. In general, the treatment of cancer involves some process by which cancer cells are destroyed while normal cells suffer less damage. Unfortunately, it isn't possible to avoid collateral damage completely, and cancer treatment via radiation or drugs is always burdened with side effects. In photodynamic therapy, the idea is to introduce a photosensitizer, a chemical that when activated by light interacts with oxygen to convert it into a very reactive form known as singlet oxygen that can destroy cells. The photosensitizer can be introduced intravenously before treating the tumor with long wavelength light via an optical fiber. Alternately, the photosensitizer can be injected directly into a tumor before the area is exposed to light. In either case, singlet oxygen is produced only within the tumor, minimizing damage to normal tissue. The process is applicable to certain types of tumors and is certainly not a cure-all for cancer.

It is this therapy that has been mangled by Antonella Carpenter, who according to investigators cheated cancer patients out of their money and gave them false hope. In spite of any evidence of her treatment having efficacy, supporters have sprung to her side, claiming that her conviction on twenty-nine counts of fraud was carried out by a kangaroo court influenced by "the greedy and vindictive genocidal maggots who control the Cancer Industry and have the FDA and courts in their back pocket." They go on to say that "the Medical Mafia is hard at work twisting the truth and vilifying Dr. Antonella Carpenter and any other non-Allopathic practitioner and natural or alternative treatments as quackery." There's more. "Dr. Antonella Carpenter was vindictively targeted by the Medical Mafia and their Gestapo goons at the FDA for successfully curing dozens of cancer patients." No. She was targeted for subjecting cancer patients to a treatment that did not work and claiming she had cured them. That is evil.

The truth is that there is no conspiracy to keep effective cancer treatments from the public. Such allegations are an insult to the thousands of researchers and physicians dedicated to solving the problem of this complex disease. As I well know, there is no real magic, only clever tricks to create the illusion that there is.

A CIRCULATING NONSENSICAL EMAIL

It's hard to know what makes people generate nonsensical emails. Perhaps it gives a feeling of power to see one's handiwork spread around the world. And that can happen, especially if the topic is cancer, and the advice given about the disease is linked to a reputable institution such as Johns Hopkins University. One such preposterous email that has really made the rounds is entitled "Johns Hopkins Update: Alternative Way to Eliminate Cancer." This diatribe certainly did not originate from Johns Hopkins, and the "advice" in it is so misleading that the university felt a need to circulate a rebuttal.

The gist of the email is that chemotherapy, radiation, and surgery are not only ineffective as treatments for cancer, but actually make the situation worse by compromising the immune system, damaging organs, and causing cancer cells to spread. A better way to cope with cancer, it said, is to starve cancer cells of the nutrients they need to multiply, create an inhospitable alkaline environment for these cells, provide enzymes to foster the growth of healthy cells, and supply the body with oxygen since cancer cells cannot thrive in an oxygenated environment. If only it were that easy. Cancer is a complex disease and there are no simple treatments or guaranteed preventative measures. While the current methods of treatment are not ideal, chemo, radiation, and surgery can effectively treat many cancers.

So, how do you "starve cancer cells"? Avoid aspartame, the widely used artificial sweetener, suggests the email, and replace it with mānuka honey or molasses. There is absolutely no evidence that cancer cells use aspartame as a nutrient, but there is evidence that cancer cells actually thrive on sugar. Indeed, one of the most interesting observations, first made by Nobel Prize winner Otto Warburg in 1924, is that cancer cells derive the energy they need to function in a fashion that is different from normal cells.

The main source of cellular energy is glucose, a simple sugar sourced from carbohydrates in the diet. In normal cells, through the process of glycolysis, glucose is converted to pyruvic acid in the liquidy part of a cell called the cytosol. The pyruvic acid then passes into the mitochondria, the cell's energy-producing organelle, where with the help of oxygen it is converted to water, carbon dioxide, and adenosine triphosphate (ATP), the molecule that serves as the cell's energy currency. This is known as aerobic respiration.

By contrast, in a cancer cell, instead of passing into the mitochondria, the pyruvic acid is converted to lactic acid in the cytosol. Since this process does not require oxygen, it is known as anaerobic respiration. It produces fewer ATP molecules per glucose, so cancer cells meet their energy needs by becoming avid consumers of glucose. Replacing artificial sweeteners with a glucose source like honey makes no sense.

The email then goes on to claim that cancer cells thrive in an acid environment, and that their multiplication can be curbed with an "alkaline diet." Yes, the production of lactic acid creates an acidic environment in the cell, but this is a result, not a cause, of anaerobic glycolysis. In any case, a cancer cell's environment cannot be altered by diet; the body does an excellent job of maintaining the blood's pH within a very narrow range, no matter what is consumed. Any suggestion that vegetables, fruits, and grains create an alkaline environment while meat is acidic has no scientific basis. Eating fruits, vegetables, and whole grains while cutting back on meat makes sense for a host of reasons, none of which involve acidity. Then there is the bit about obtaining live enzymes by eating raw vegetables. Enzymes cannot be alive or dead; they are just special proteins that the body uses as catalysts — that is, substances that speed up chemical reactions. But enzymes consumed orally are digested like any other protein.

Finally, there is the claim that cancer cells cannot thrive in an oxygenated environment, and that exercising and deep breathing help to get more oxygen down to the cellular level. Once more, a seed of truth is nurtured into a jungle of twaddle. Indeed, aerobic respiration, the desired metabolic process, requires oxygen and anaerobic glycolysis does not. But cancer cells do not switch to anaerobic glycolysis because of a lack of oxygen and cannot be goaded into aerobic respiration by increasing oxygen. In any case, the bloodstream is normally saturated with oxygen, although perhaps not quite to the extent that this "Johns Hopkins email" is saturated with nonsense.

SPOON-BENDING FIASCO

Everyone has skeletons in their closet. There's at least one in mine. A couple of years ago while on a cruise, I pinched a spoon from the dining room. It wasn't because of any lack of spoons at home. It was because no matter how hard I tried, I could not bend this one. I tried with two hands. I tried by pushing against the table. I even tried placing the handle under my heel and tugging on the head. No give at all. I had to have that spoon!

I've been practicing magic as a hobby ever since I was a teenager. It has turned out to be a perfect fit with my career because of the numerous scientific principles involved in creating the illusion of contravening the laws of nature. That's what magic is all about! Seeing someone levitate, or vanish inside a cabinet, or appear out of thin air, requires an apparent suspension of the laws of nature. The key word of course is "apparent," because all such effects are accomplished by clever scientific means. A magician, however, attempts to ensure that the audience will not discover those means. Science can also appear magical, but in this case, we relish in scuttling the magic with down-to-earth explanations. Just think about it. Isn't an airplane with hundreds of people aboard flying through the air magical? How about taking pictures with your smartphone and sending them around the world in seconds? Or a seed growing into a plant? Or a new life being created from the meeting of cells? But magic is converted into science with an appropriate explanation.

I have found performing magic to be an excellent springboard for a discussion of scientific methodology and for fostering the critical thinking needed to prevent being swept away by the tsunami of pseudoscience generated by a rapidly multiplying bevy of charlatans. When you can demonstrate how "psychic surgery," a procedure by which diseased tissues are apparently removed without an incision, can actually be accomplished by sleight of hand, you have given believers something to think about. Similarly, a demonstration of "mental" effects, with a clear declaration that these are done by clever chicanery, can help convince at least some that trickery may be involved when psychics perform seemingly scientifically inexplicable feats.

One such feat is "psychokinesis," or the ability to move objects using only the power of the mind. Psychokinetic effects were first popularized in the middle of the nineteenth century when Angelique Cottin in France claimed that electric emanations from her body allowed her to move objects without touching them. She convinced many observers of her power, but critics offered down-to-earth explanations about how such effects could be performed by trickery. Since that time, numerous psychics have claimed psychokinetic powers, with Uri Geller being perhaps the most famous. In the 1970s, he beguiled audiences and even some scientists with his apparent ability to bend metal with the power of his mind. He gets credit for introducing the phenomenon of mental spoon bending, an exhibit upon which he built quite a spectacular career.

Magicians were also astounded. Not by the trick, which can be accomplished by a number of established methods, but by how the public was so ready to swallow a "paranormal" explanation. Conjurers were quick to reproduce the spoon bending trick, pointing out that the only requirement was a modicum of sleight of hand. This brings us back to my pilfered spoon.

When I do the spoon-bending trick, I first hand out the spoon to the audience with a challenge to bend it. Once it is established that it can withstand all efforts, I proceed to bend it "with the power of my mind." But in rare cases, some strong men have managed to bend the spoon and destroy my performance, so I'm always on the lookout for super-strong spoons. I can tell you that Crystal Cruises have such spoons. They absolutely cannot be bent, except in the hands of a magician who is equipped with a "special something."

But why am I talking about tormenting cutlery? Because last week, thanks to colleague Tim Caulfield, a health law professor at the University of Alberta, I learned that "Integrative Pediatric Medicine Rounds" at his university were set to feature a talk on "Spoon Bending and the Power of the Mind." The seminar would be given by an "energy healer," who was described as being "a Reiki Master teacher, a certified Trilotherapy practitioner, a Yuen Method practitioner, and a teacher of popular Spoon Bending and Tantric Sex workshops." So this was not to be a workshop on critical thinking, which could have been appropriate. The prospective speaker actually claimed that 75 percent of attendees would be able to bend spoons with their mental energy!

The scientific community reacted with vigor to this assault on reason, and the resulting extensive media coverage caused the seminar to be canceled with some weasel explanations being provided about the workshop "being withdrawn by the presenters."

The "presenter" was to be Anastasia Kutt, who is not some wacky outsider, but is listed in the university's directory as "a research assistant" in the Complementary and Alternative Research and Education (CARE) Program who is also involved in "research activities and organizing events." What sort of events? Given her interest in topics such as tantric sex and spoon bending, one wonders.

Criticism of this spoon-bending fiasco should not be construed as an attempt by the mainstream scientific community to curb free speech or to police academic research. Rather it is an appeal for reason and for vigilance against quackery sneaking into "integrative medicine" programs, which are becoming increasingly popular.

I don't know how Ms. Kutt bends spoons, but I'd be willing to fly to Edmonton at my expense to find out. If she can bend my Crystal Cruises spoon, I'll eat a University of Alberta Integrative Health Program hat.

A HOUDINI LOW POINT

National Magic Day is celebrated annually on October 31 in memory of Harry Houdini, who passed away on that day in 1926. While his name has become synonymous with magic and escapes, it should be remembered that Houdini was also a prolific writer, authoring such classics as *The Right Way to Do Wrong*, *A Magician among the Spirits*, and *Miracle Mongers and Their Methods*. But his most curious work, published in 1909, was *The Unmasking of Robert-Houdin*, in which he aims to topple his former hero from his pedestal by showing that the illusions and tricks that the celebrated magician claimed were his invention were actually stolen from others.

The book concludes with the following vicious attack on the performer whose name young Ehrich Weiss pinched to become "Houdini": "The master-magician, unmasked, stands forth in all his hideous nakedness of historical proof, the prince of pilferers. That he might bask for a few hours in public adulation, he purloined the ideas of magicians long dead and buried, and proclaimed these as the fruits of his own inventive genius. That he might be known to posterity as the king of conjurers, he sold his birthright of manhood and honor for a mere mess of pottage, his *Memoirs*, written by the hand of another man, who at his instigation belittled his contemporaries, and juggled facts and truth to further his egotistical, jealous ambitions."

What makes this really curious is that "juggled facts and truth to further his egotistical, jealous ambitions" can well be applied to Houdini himself. For example, he always maintained that he had been born in Appleton, Wisconsin, when he had actually been born in Hungary, believing that being born in America would be more appealing to his audiences. Houdini's egotism was legendary; he vigorously attacked other magicians he believed were imitating his performances. It was partly an attempt to discredit anyone he viewed as an imitator that spawned his attack on Robert-Houdin. The other contributing factor was Robert-Houdin's widow expressing no desire to meet him when he traveled to France to walk in the footsteps of his hero. He couldn't accept that the great Houdini could be spurned in such a fashion and proceeded to lash out with a venomous assault on Robert-Houdin's heritage, which he claimed was undeserved.

Houdini proceeded to demonstrate how Robert-Houdin's most famous effects such as the Orange Tree and the Ethereal Suspension were not original. But Robert-Houdin never claimed they were; he referred to them as new tricks, which indeed they were. The truth is that virtually all novel magic effects are modifications of older tricks. What matters is the performance, and Robert-Houdin's was novel.

His Ethereal Suspension was stimulated by the discovery of ether anesthesia in Boston in 1846. The magician pretended to put his son to sleep with ether and then, defying gravity, suspended the boy parallel to the ground with his elbow resting on a rod, apparently the sole support. Indeed, the effect was not new; street artists in India had been performing it since the 1830s. The secret relies on the rod being inserted into a support hidden under the performer's clothes. These days you can see the effect staged in public squares around the world with tourists gawking at the entertainer apparently floating in air. But Robert-Houdin's linking the performance to the discovery of ether was not only novel, it introduced audiences to the concept of anesthesia.

The Orange Tree was Robert-Houdin's signature effect that enchanted the audience at his theater in Paris. He vanished a ring inside a handkerchief and then introduced an orange tree sporting only green leaves. Spontaneously the tree sprouted flowers that turned into real oranges, which were handed out to the audience. One orange stayed on the tree, opening up to reveal two fluttering butterflies that lifted a handkerchief bearing the previously vanished ring. The tree was a mechanical marvel, made by Robert-Houdin, who had originally been trained as a clockmaker. Certainly automata driven by various types of clockwork mechanisms had been around for centuries. Prague's famous astronomical clock created in the fifteenth century featured automata portraying the twelve apostles and the figure of death in the form of a skeleton that struck a bell to ring out the time, a reminder of how short life is and the need to use your time well. It still works today. Around the same time Leonardo da Vinci displayed a mechanical knight that could stand, sit, and maneuver its arms through a system of pulleys and cables. Robert-Houdin added to the genius of those who came before to design his marvelous orange tree, a replica of which still amazes audiences today.

Houdini was a superb performer and great opponent of pseudoscience and his memory deserves to be celebrated every October 31. His attempt to elevate himself by deflating the accomplishments of his famous predecessor is an unfortunate blight on a stellar career.

TORNADOES, RAINBOWS, AND CHEMISTRY

Moses, with a little help from above, proceeded to part the waters as a pillar of fire barred the pharaoh's soldiers from pursuing the Hebrews. A truly memorable scene from Cecil B. DeMille's 1956 classic, *The Ten Commandments*. In the film, the fire was the clever handiwork of animators, but it could have been made more realistic with a little help from chemistry. I'm thinking of a classic demonstration known as the methanol tornado. But before we get into this, let me repeat one of my mantras: "There are no safe or dangerous chemicals, only safe or dangerous ways to use them." And methanol is a classic example. This simple alcohol is not only extremely flammable, it is highly toxic through ingestion or even skin exposure. That, though, does not mean that it cannot be used instructionally with the proper precautions.

The demonstration requires a metal mesh garbage can, some sort of platform that can be made to spin, a small glass dish with a few milliliters of methanol, a chemical that can color a flame, and a source of ignition. The waste basket is placed on the spinning platform with the methanol container at the bottom. Spinning the platform after igniting the methanol causes a pillar of fire to rise in an impressive fashion and the addition of a bit of strontium nitrate to the methanol produces a striking red color. Seems to me that a close up of this fire tornado can make for a great movie special effect, but when you play with fire, as the saying goes, you can get burned. That's just what happened to eight children and an adult in a science museum in Reno, Nevada.

Exactly what occurred isn't clear, but the basic problem was that the stock bottle of methanol, from which a little had been poured out for the experiment, was left on the demonstration table. Somehow, when the methanol in the experimental dish was being ignited, the invisible vapors from the bottle also caught fire. This then ignited the methanol in the bottle, causing it to tip over and spill its burning contents onto the floor where the spectators were sitting. This demonstration is performed routinely in science museums and educational institutions around the world with no problem, but the glitch here was the stock bottle of methanol that was left nearby. That is an emphatic no-no!

While the accident with the methanol tornado experiment was an extremely rare occurrence, another demonstration with methanol, called the Rainbow Flame Experiment, has caused numerous accidents, injuring students and teachers when appropriate safety procedures are not followed, usually through a teacher's ignorance. The experiment is a very popular one because it teaches students how to test for the presence of certain metals by virtue of the color they impart to a flame. The most common way of carrying out the experiment involves setting up a series of small glass dishes with each one containing a tiny amount of methanol in which a metal salt has been dissolved. When the methanol-metal samples are ignited, a rainbow of colors is produced. A salt that has sodium produces a yellow color, copper produces blue, boron green, potassium lilac, lithium carmine, and strontium red.

The principle here is that the heat from the flame excites electrons in the metal ion to higher energy states, and when they drop down to lower energy levels they emit the energy they had absorbed in the form of light. Since the difference between energy levels is a property of the atoms in question, the specific wavelengths of light emitted, in other words the colors, can be used to identify the presence of a metal. The brilliant colors of fireworks, the bright red strontium flame of an emergency roadside flare, and the yellow glow of a sodium vapor highway lamp are all due to such electronic transitions.

The accidents while performing the rainbow experiment have occurred when a container of methanol that should never have been anywhere in the vicinity ignited. Commonly, the experimenter attempted to revitalize the flame by pouring more methanol into the dish, setting the stream on fire and creating a flashback into the bottle. This is similar to the disaster that can take place when someone squirts lighter fluid onto a lit barbecue and ends up barbecuing themselves.

The frightening number of rainbow accidents has led chemical safety experts to recommend that using methanol as the source of fuel be abandoned. I would rather see educators be properly educated in carrying out such demos, and to have them emphasize to the students that the methanol bottle must be returned to its appropriate place of storage before proceeding with the experiment. It is true, however, that the chemical principle involved can be demonstrated without methanol. Just place a small amount of the salt in question on a platinum, gold, or silver wire that is then passed through a Bunsen burner. These metals do not produce a color of their own so any observed change is due to the sample being investigated.

There is another good reason to perform the experiment with a Bunsen burner. Students can be told the fascinating story of how Robert Bunsen actually developed the iconic burner in order to study the colors produced by metals in its flame. Isaac Newton had previously shown how a prism can be used to separate white light into the colors of the rainbow, and Bunsen applied this principle to separate the colors of a flame into their individual components. Together with the physicist Gustav Kirchhoff, he developed the spectroscope, an instrument still used today to identify unknown substances by the colors they produce when heated. The composition of stars is actually determined by examining the light they emit through a spectroscope. A heavenly invention one might say.

HIJACKING CHEMISTRY

A picture is worth a thousand words. And a live demo is worth a thousand pictures. That rings especially true when it comes to chemistry. To this day, I remember that epic moment in introductory physical chemistry when the prof combined two colorless solutions and within a few seconds the mixture flashed to a dark blue color. The whole class gasped! Alas, it was the only demo I witnessed in my whole undergraduate career. As I was to learn, I had been thrilled by the classic "iodine clock reaction," so named because the reagents can be adjusted so that the time it takes for the color change to occur can be quite accurately predicted.

The essence of this reaction is the combination of iodine with starch to form a deep blue color. Many a student has been impressed by cutting a potato in half and turning it blue with tincture of iodine. Indeed, an iodine solution is a common way to test for the presence of starch. For example, as an apple ripens, its sweetness increases as starch converts into simple sugars. Farmers can therefore check whether an apple is ripe by cutting it in half and spraying the surface with iodine. If the iodine causes an intense blue color, the apple isn't ripe. In this instance, the color formation is practically instantaneous. Some clever chemistry is needed to produce a time delay.

The clock reaction starts with the ionic form of iodine, namely the iodide ion, which does not react with starch. But iodide does react with hydrogen peroxide in an oxidation reaction to produce iodine. Now, here is the ingenious part: vitamin C is a well-known antioxidant and will convert iodine back to iodide. If it is combined with iodide and added to a solution of starch and hydrogen peroxide, it will immediately react with the iodine that is formed and prevent it from reacting with the starch. But as soon as all the vitamin C is used up, any newly formed iodine will combine with starch to produce the blue color.

I was so impressed by that reaction that I scoured the library for books on chemical demonstrations and came across *Tested Demonstrations in Chemistry* by Hubert Alyea. That turned out to be a life-changing moment. The book was full of fascinating demos that I would eventually perform in class. Alyea, I learned, was a professor of chemistry at Princeton who had become famous for his public lectures that featured colorful and sometimes explosive demonstrations, along with solid scientific explanations mixed with some zany prattle. He had earned the nickname Dr. Boom, and had served as the inspiration for Fred MacMurray's character in Disney's *The Absent-Minded Professor*, even meeting with the actor so he could study his mannerisms.

As you might imagine, when I heard that Alyea was going to give one of his presentations in Montreal, I didn't walk, I ran. I wasn't disappointed. The *New York Times* had described his performance accurately: "Amid explosions and swishing clouds of carbon dioxide, Alyea explains the mysteries of chemistry with contagious enthusiasm . . . with the drama and verve of a sound-and-light show." He even upped the clock reaction I had already become familiar with! In his hands, not only did the solution change from colorless to blue, it reverted to colorless, only to repeat the cycle again. I was totally taken by this oscillating reaction and years later made it a focal point of "The Magic of Chemistry," a show my colleagues and I have performed across North America. I think given his passion for bringing chemistry to the public, Dr. Alyea would have approved.

But I think he would have had some snarky comments about a scheme using the iodine-starch reaction in China. Tea of course is a big deal over there. Remember the expression "for all the tea in China"? So it isn't surprising that it is common for tourist groups to make a stop at a tea plantation where they hear all about the wonders of drinking tea before being ushered into a shop where they can deposit their Western currency. The talk on the benefits of tea is complemented by a neat little chemical demonstration. Rice is placed into a bowl and is mashed with a pestle. Water is added, followed by a solution containing iodine, although it is not identified as such. Rice, like potato, contains starch, and presto, we have the formation of the characteristic blue color.

Then comes the kicker. Freshly brewed tea is added to the blue rice solution, instantly eliminating the color. As this happens, the group is treated to a discussion of antioxidants and how such substances eliminate the free radicals that are associated with a variety of diseases as well as aging. Next comes a description of all the antioxidants that are found in tea with the implication that their presence is being demonstrated by the vanishing of the blue color. The underlying message is that tea can clean the inside of our body the same way that it has cleansed the rice solution.

The demo is quite compelling and sends the onlookers scurrying to the tea shop. But if they scurried towards a chemistry book instead, they would find a different explanation for the color change. Tea does contain a number of polyphenols in the catechin category that are indeed antioxidants. But the color change noted here has nothing to do with their antioxidant effect. Rather it has to do with a reaction that organic chemistry students will recognize as electrophilic aromatic substitution. Basically, iodine substitutes for a hydrogen atom in the polyphenols. This means that the iodine is used up, so there isn't any left to complex with starch to produce the blue color. There is absolutely nothing wrong with drinking tea; its antioxidants may actually be beneficial. But the discoloration of the rice and iodine mixture does not demonstrate that. My favorite demo has been hijacked!

A TOXIC CLEANSE

Picture this. You swallow a little pill, wait until it irritates your intestine enough to expel its contents, and then you hunt through the expelled excrement to retrieve the pill. Why? So you can use it again next time you need to get rid of the bad humors in your body that are making you sick. How can a pill survive passage through the digestive tract? It can, if it is made of metal — in this case, antimony. Now, don't go asking the pharmacist for antimony pills. The scenario just described isn't current; it was plucked out of the Middle Ages when the cure for disease was to expel "bad humors" from the body. Actually, that was not unlike the current craze of expelling unnamed toxins from the body with a variety of "cleanses," many of which have a laxative effect.

Hopefully nobody today would be silly enough to use antimony or its compounds, because these are truly toxic. Of course, they didn't realize that in the Middle Ages; all they knew was that antimony was pretty good at evacuating the body. And not only through the rear portals. One method involved drinking wine that had been left standing overnight in a cup made of antimony. This resulted in the antimony reacting with tartaric acid in the wine to form antimony tartrate, a compound that induces vomiting. The idea of purging the body to treat illness persisted into the late stages of the eighteenth century. When Mozart came down with a mysterious illness, he was treated with tartar emetic, as antimony tartrate was commonly called. What ailment he suffered from isn't clear, but he died within two weeks. His symptoms of intense vomiting, fever, swollen abdomen, and swollen limbs are consistent with antimony poisoning. Of course, we cannot prove that antimony was responsible for Mozart's death; he also suffered from rheumatic fever since childhood, a condition that may have led to his demise at a young age.

Mozart had always been sickly and it is well-known that he had been often treated with antimony compounds by his physicians and that he even dosed himself when he didn't feel well. It is interesting that Mozart actually believed he was being poisoned, but not by himself. He thought his musical rival Antonio Salieri was trying to do him in. Although the famous movie *Amadeus* alludes to this possibility, historical facts do not corroborate the poisoning story. Contrary to the portrayal, Salieri did not confess at the end of his life to having tried to kill Mozart.

Back in the 1990s, a volatile compound of antimony known as stibine (SbH_3) was accused of being responsible for crib death. The theory was that it was produced from antimony oxide added as a flame retardant to polyvinyl chloride sheets. A fungus found in mattresses supposedly made this conversion possible, at least under laboratory conditions. The theory has now been dismissed because neither the fungus, nor levels of antimony in babies' blood could be correlated with crib death.

More recently Greenpeace created a stir with a booklet entitled *A Little Story about the Monsters in Your Closet*. What sort of "monsters"? The subtitle brings them out of the closet: "Study Finds Hazardous Chemicals in Children's Clothing." Yup, the monsters are chemicals. One chemical that the Greenpeace study detected was antimony trioxide, present in all fabrics that have polyester as a component. No great surprise here since antimony trioxide is used as a catalyst in the production of polyester as well as a flame retardant. And it is true that antimony trioxide can be described as presenting a hazard. But hazard is not the same as risk.

Hazard is the innate potential of a substance to cause harm without taking into account extent or type of exposure. Inhalation of antimony compounds in an occupational setting can be a problem, and it is correct that antimony trioxide has been classified as "suspected of causing cancer via inhalation." But this is not relevant for the trace amounts found in fabrics. Here the issue would be migration out of the fabric and subsequent absorption. This has been extensively investigated and the amounts that are encountered are well below the established migration limits. The same applies to the trace amounts that leach out of the polyester bottles that are widely used for water and other beverages. Concentrations are less than the 5 parts per billion safety limit.

Antimony does not occur in nature in its metallic form, so where did Middle Age physicians get it? Like most metals, antimony has to be smelted from its ore, in this case antimony sulfide, also known as stibnite, a substance that has been known for thousands of years. Jezebel, the Biblical temptress, is said to have used it to darken her eyebrows and stibnite was the main ingredient in kohl, used by ancient Egyptian women in a type of mascara. Exactly who figured out that heating antimony sulfide converts it to antimony oxide, which yields metallic antimony when fired with carbon, is unknown, but if you visit the

Louvre, you can see a 5000-year-old vase that is made of almost pure antimony.

Today, neither metallic antimony nor its compounds have a medical use, although up to the 1970s, antimony compounds were used to treat parasitic infections like schistosomiasis. These preparations did kill the parasites, but sometimes they also dispatched the patient. Up to the early twentieth century, tartar emetic was used as a remedy, albeit an ineffective one, for alcohol abuse. The *New England Journal of Medicine* once reported a case of a man whose wife tried to cure him of his alcoholic habit by secretly putting tartar emetic into his orange juice. The result was a trip to the hospital with chest pains and liver toxicity. Two years later, the man reported complete abstinence from alcohol. Seems antimony had taught him a lesson.

THE REAL FLINTSTONES

I grew up watching *The Flintstones*, the first animated television sitcom. Watching the antics of that modern Stone Age family was a lot of fun, but it wasn't until one of our rare ventures into the chemistry lab in high school that I gave a thought to the family's name. Mr. Cook showed us how to light a Bunsen burner with a "flintstone," which, as he told us, was the way man first availed himself of fire. That provided a rationale for the choice of the family name, although the story about the flintstone being the first "fire starter" was incorrect.

There is much debate about humankind's first controlled use of fire, with estimates ranging from some 400,000 to a couple of million years ago. Producing the first practical flame likely involved carrying glowing embers from a natural fire started by lightning or by a volcanic eruption, keeping these smoldering by adding wood or animal fat, and finally adding dry tinder and blowing on the embers. The first actual creation of fire likely relied on hand-spinning a stick against another piece of wood to generate dust that would ignite by the heat produced by friction.

Then sometime during the Stone Age, upon fashioning tools out of stone, our ancestors noted that striking certain rocks against each other produced sparks. Flint is a variety of quartz, composed mainly of silicon dioxide. It can sport a variety of colors due to the inclusion of other minerals and is almost as hard as diamond. Striking flint against an iron-containing rock, such as pyrites, will make sparks fly! These are glowing bits of iron that can start a fire on contact with easily combustible materials.

The technology was greatly improved during the Iron Age when the heating of iron ore with charcoal yielded metallic iron, convertible into steel by alloying with carbon. Sparks were produced much more readily when steel was struck against flint. Until the introduction of matches in the nineteenth century, the use of flintstones was the most common way to spark a fire. People carried tinderboxes, which held a flint, a piece of steel, some dry matter such as scorched cloth or dried fungus, and a wooden splint tipped with sulfur. With a little experience, striking the flint with the steel and blowing the sparks onto the tinder produced a flame that would then ignite the splint, which was used to light a candle or kerosene lamp.

Early firearms featured a flintlock mechanism with a piece of flint attached to a spring-loaded hammer that was released when the trigger was pulled. The flint struck a piece of steel, sending a shower of sparks into a small pan of gunpowder that ignited and touched off the main charge that had been packed into the barrel.

Although matches replaced tinderboxes, flintstones did not

disappear. They are still used in cigarette lighters, and in strikers to safely light welding torches, and, yes, in Bunsen burners. But what we now call flint isn't really flint. It is a man-made material called ferrocerium, composed of iron, cerium, lanthanum, and magnesium. Actually, it takes on the role of the steel, not the flint. With traditional flint lighters, it is small particles of steel produced by striking against flint that ignite. In this case, thanks to cerium's low ignition temperature, it is particles of the ferrocerium "flint" that are ignited when struck with a hard metal.

Today, such "flint" lighters are often included in emergency survival kits. A strip of ferrocerium is glued onto a small block of magnesium, a soft metal that burns with a hot flame. The technique is to use a knife to shave off a small pile of magnesium flakes and then ignite these with sparks produced by rubbing the knife on the ferrocerium strip. The advantage of this system is that, unlike with matches, wet magnesium will still burn. The bright light produced by burning magnesium came in handy during the early days of photography when film was very insensitive. Special lamps that burned strips of magnesium came to be commonly used.

Large pieces of magnesium are actually difficult to ignite, so there is no danger of a whole magnesium block bursting into flame. But once magnesium does catch fire, it is extremely difficult to extinguish, burning even in the absence of oxygen, since magnesium reacts with nitrogen in the air to form magnesium nitride. Dousing with water just makes the situation worse because the metal reacts with water to produce combustible hydrogen gas. That makes magnesium useful for the construction of incendiary bombs. During World War II, German cities were engulfed by firestorms triggered by two-kilogram magnesium bombs set off by a thermite reaction within the bomb. Powdered aluminum when combined with iron oxide produces molten iron and a huge amount of heat, enough to ignite the magnesium casing. Once aflame, the bombs set fire to everything in their path.

Where does magnesium come from? It is widely found in nature, although never in its elemental form. It is at the heart of chlorophyll, the molecule that enables the capture of energy from the sun needed to convert carbon dioxide and water into glucose, which is then used by the plant to make myriad important compounds ranging from starch to DNA. Plants get the magnesium from the soil, where it occurs in the form of numerous minerals such as magnesium carbonate. This mineral is also the source of much of the magnesium metal produced in the world. This is done by first heating the magnesium carbonate to produce magnesium oxide, which can then be reduced to simple magnesium by reacting it with silicon. Seawater also contains vast amounts of extractable magnesium chloride. Passing an electric current through molten magnesium chloride yields magnesium at the cathode and chlorine at the anode, basically the same process used by Humphry Davy in 1808 to isolate the first pure sample of magnesium.

Think of all this the next time you watch *The Flintstones*. And please ignore the folly of humans and dinosaurs coexisting.

TESLA: A SPARK OF GENIUS

I don't claim it was any sort of a scientific survey, but the results are interesting nevertheless. I polled a pretty diverse cross section of people about Nikola Tesla. Some thought he was the inventor of the electric car; a few students, remembering that the tesla is a unit used to measure magnetic flux density, connected him ambiguously to magnetism; but the majority of those asked had no idea that most of our modern conveniences can be traced to the brilliant mind of the Serbian scientist who moved to America in 1884. Our well-lit homes, vacuum cleaners, refrigerators, washing machines, and electric garage door openers can all be traced back to Tesla's inventions, which made the transmission of electricity practical and the electric motor functional.

Tesla's alternating current could be transmitted easily by wire over great distances while its competitor, Edison's direct current, required power generators every few miles. The "battle of currents" was fierce, speckled with some bizarre demonstrations. Edison electrocuted animals with alternating current to show how dangerous it was while Tesla subjected himself to 25,000 volt jolts to show the safety of his technology.

When he first came to America, Tesla actually worked for Edison and took up the challenge of solving some of the inventor's problems with direct current. He was offered $50,000 if he came up with a solution. He did and laid claim to the money. Edison responded, saying, "Tesla, you don't understand American humor." But Tesla had the last laugh because it was his alternating current that eventually powered the world.

How did he get interested in electricity? It was thanks to his pet cat, Macak, Tesla claimed. When he was a boy he saw sparks fly on stroking the cat's fur, which seemed to him like miniature lightning bolts. Is nature a gigantic cat? he wondered. The curiosity aroused by the cat eventually led to the invention of the Tesla coil, capable of generating huge voltages and producing the largest man-made lightning bolts ever seen.

The inventor was obsessed with the transmission of energy without wires and was captivated by vibrational frequencies. After all, sound waves of the right frequency could shatter a glass! In his New York lab, he built a vibrating platform powered by compressed air to see how objects reacted to vibrations of different frequencies. One day, as he stepped onto his platform while it was in action, he noted "a strange but pleasant feeling." Then, as he stood enjoying himself, another effect emerged. He had invented a mechanical laxative! Soon Mark Twain would benefit from that invention. Sort of.

Tesla had been an admirer of the American writer since he was a child. Afflicted with cholera, young Nikola was bedridden when a nurse brought him a Serbian translation of Twain's works. He was smitten with the stories of the American South and was absolutely thrilled to meet the author when both were living in New York.

One day when Twain was visiting Tesla's laboratory, he complained about his constant bouts with constipation. The inventor suggested that the vibrations produced by the oscillating plate of his generator might "impart vitality." Twain was quite scientifically minded and was game for the experiment, so Tesla revved up the machine and warned Twain that he must step off the platform when told. Twain apparently enjoyed the vibrations running through his body to such an extent that he ignored Tesla's advice to step off. Soon the lab reverberated with a panicked cry: "Tesla! The water closet! Where is it?" Apparently the dash towards the toilet was too late. The humorist's signature white suit was no longer quite as white as it had been.

According to Tesla, his oscillator was capable of doing more than just scaring the poop out of people. It could even improve ladies' complexion, as Tesla believed feminine beauty had a direct relationship to bowel movements. Furthermore, his device could cause buildings to resonate. Tesla even claimed that one of his experiments triggered an earthquake in the city. That was questionable, but there was no doubt about his lighting a bank of bulbs in his Colorado Springs laboratory with no wires attached. Exactly how he did it isn't known because Tesla was very secretive and didn't keep proper written records. He also built a 185-foot-high transmission tower at his Wardenclyffe laboratory on Long Island, topped with a giant copper dome. He planned to transmit free unlimited wireless electricity all over the world, but his backer J. Pierpont Morgan lost confidence in the project and stopped funding it. Today, the wireless transfer of energy as first demonstrated by Tesla is being realized. Just think of wireless chargers for toothbrushes and cell phones.

And that is still not all Tesla did. He introduced the concept of radio two years before Guglielmo Marconi — who received the Nobel Prize for the discovery. Tesla's comment was "Marconi is a good fellow. Let him continue. He is using seventeen of my patents." Tesla also engineered the first hydroelectric plant at Niagara Falls, produced ball lightning in his laboratory, made X-ray images before Wilhelm Röntgen and envisioned radar decades before it was "invented."

Like many geniuses, Tesla was also somewhat eccentric. He could not stand to touch hair and was revolted by the sight of pearls. In his later years, he became obsessed with feeding pigeons, describing his relationship with one particular bird as "loving her as a man loves a woman." But idiosyncrasies aside, no scientist has had a bigger impact on our life than Nikola Tesla. It's a strange world where the Kardashians are famous for having done nothing, while Tesla, at least as far as the general public is concerned, remains in relative obscurity despite having been the major architect of the electronic age.

SULFUR'S COLORFUL PAST

Looking across the harbor from Vancouver, you can't miss the giant mounds of sulfur waiting to be shipped around the world, mostly destined to be converted into sulfuric acid, one of the most important industrial chemicals in the world.

While Canada does not have any sulfur mines, we do have huge amounts of natural gas and a significant petroleum industry. Natural gas is roughly 95 percent methane but is often "soured" with hydrogen sulfide, which has the notorious odor of rotten eggs. During refining, hydrogen sulfide is separated and is then reacted with oxygen to yield elemental sulfur. Crude oil can also yield sulfur. While it is mostly a complex mix of hydrocarbons, the oil also contains some hydrogen sulfide and close to 4,000 different sulfur compounds. These can be separated from the oil, converted to hydrogen sulfide and then to sulfur. In Canada, most of the sulfur recovery from sour gas occurs in Alberta and some in British Columbia. The tar sands in Alberta also yield sulfur as do oil refineries in Eastern Canada.

There is another benefit to the removal of sulfur compounds from sour gas and petroleum, other than providing the raw material needed for the synthesis of sulfuric acid. When fuel containing sulfur compounds is burned, sulfur dioxide is produced. This gas reacts with moisture in the atmosphere to yield sulfuric acid, resulting in acid rain.

The production of sulfuric acid from sulfur was first commercialized in 1763 by, of all people, Dr. Joshua Ward, an English physician of questionable repute. He scaled up a chemical reaction discovered by Johann Glauber, a German-Dutch chemist, who had shown that heating sulfur with steam and potassium nitrate, commonly known as saltpeter, yielded sulfuric acid. Saltpeter releases oxygen, which oxidizes the sulfur to sulfur trioxide, which then reacts with water to form the acid.

Ward was an interesting character who referred to himself as "The Restorer of Health, and Father to the Poor." Other physicians had a different opinion. They ridiculed Joshua Ward's

Drop, a medicine that Ward had invented and modestly named after himself. It was supposed to cure people of any illness they had. He was also the inventor of Friar's Balsam, a concoction made from a tree resin that was said to be useful for breathing problems, coughs, laryngitis, as well as skin abrasions and lesions. It is still sold today. While Ward's medical knowledge was questionable, his chemistry was sound. The modern production of sulfuric acid still relies on the principle of converting sulfur to sulfur trioxide and then reacting this gas with water.

Sulfuric acid is mainly used to feed the world through the production of phosphate fertilizers. Various types of phosphate rock (fluoroapatite) are treated with sulfuric acid to yield phosphoric acid, which can then be reacted with ammonia to produce ammonium phosphate, a common fertilizer. Sulfuric acid also finds use in the production of detergents, plastics, dyes, insecticides, batteries, inks, lubricants, textiles, and explosives. That's why it has been said that a nation's production of sulfuric acid is an indicator of its industrial strength.

The first substances known to humans in their elemental form were gold and sulfur. The bible speaks of fire and brimstone and the Ebers Papyrus dating from 1550 B.C. describes an eye salve containing sulfur. The ancients knew sulfur as the "burning stone" and undoubtedly were fascinated by the bluish flame and pungent odor it produced. The Greeks burned sulfur to purify their temples, and that led to an accidental discovery that proved to be very useful. Rodents perished when exposed to sulfur dioxide vapors! This introduced the idea of using burning sulfur as a fumigant to kill pests. In the *Odyssey*, Homer talks about burning sulfur to preserve corpses in the hot sun.

The ancient Romans recognized the ability of the sulfur dioxide formed by burning sulfur to bleach wool and, according to some accounts, to kill rogue microbes in wine barrels, although there is controversy about whether they actually did this. We do know for sure that by the fifteenth century, sulfur candles were burnt inside barrels before filling them to help preserve the wine. The Greek philosopher Theophrastus described how rubbing cinnabar, or mercury sulfide, with vinegar in a copper bowl produced pure mercury. This may well have been the first chemical reaction ever recorded. Turning a red powder into shiny metallic mercury must have seemed absolutely magical! It was such reactions that gave rise to the alchemical quest to produce gold via chemical processes. Sulfur's yellow glow and mercury's metallic gleam made these two elements the focal points for the alchemists' ill-fated attempts.

By the thirteenth century, sulfur's combustibility began to play a different role. A powder produced by combining saltpeter with charcoal and sulfur was found to ignite easily. It came to be known as gunpowder since the quickly expanding gases produced on combustion could be used to launch various projectiles. Gunpowder would change the world! Sulfur went from being an esoteric substance used by alchemists to a highly desired commodity for warfare. And since warfare often pitted Christians against "infidels," the Church was determined to keep gunpowder out of the hands of its enemies. In 1527, Pope Clement VII issued an edict to excommunicate anyone who traded sulfur to "Saracens, Turks, and other enemies of the Christian name."

Venezuelan president Hugo Chavez had a different take on sulfur. In 2006, when speaking at the United Nations the day after George Bush had spoken at the same podium, he commented that the "devil spoke here yesterday" and "it smells of sulfur still today." Compounds of sulfur can have a nasty smell. Just think of skunk, rotten eggs, or flatus. Elemental sulfur, however, is a yellow solid that has no odor. Chavez should have stuck to politics because he didn't get his chemistry right.

SEEING THROUGH THE SMOKE

The picture of the kerosene lamp a "friend" posted on my Facebook page was intriguing. Its ornate features smacked of the Victorian era, but what made it particularly noteworthy was the metal bowl suspended above the open-ended glass cylinder that surrounded the wick. What was this lamp supposed to do, the friend queried. And that question immediately transported me back to my youth.

As a child, I suffered from asthma, and back then, in those pre–rescue inhaler days, there was a treatment that involved placing a powder in a metal cup, holding it above a flame, and breathing in the fumes that were released. Of course, I was too young to wonder about what was going on. I just did as I was told. Happily, the treatment worked! However, with the appearance of metered inhalers dispensing isoproterenol, the powder regimen fell by the wayside. And truthfully, I hadn't given it much thought until the question about this lamp jarred my memory. It certainly looked like it was designed to vaporize some substance placed in the bowl above the flame.

Thanks to Google, it didn't take long to identify the lamp as a Vapo-Cresolene inhaler, first produced in 1879. As I had suspected, the intent was to release a vapor that was claimed to cure respiratory conditions such as asthma and whooping cough. The device was also promoted as having "powerful germ destroying properties."

Louis Pasteur's recent discovery of disease-causing "germs" had made it into the popular press, as had British surgeon Joseph Lister's championing the use of carbolic acid (phenol) to kill such germs during surgery. Lister had noted that surgical wounds often "festered" and produced an obnoxious smell that was a prelude to disaster. Noting that carbolic acid was often spread on fields to counter the stench produced by fertilizing with sewage and that animals later grazing on these fields suffered no ill effects, Lister began to treat wounds with carbolic acid. He also designed a sprayer to eliminate germs in the operating room, earning him a reputation as "the father of modern surgery."

The Vapo-Cresolene lamp attempted to capitalize on Pasteur's and Lister's work. At the time, chemists were focused on the large variety of chemicals derived from coal tar, including a distillate known as creosote. It was from creosote that Lister's carbolic acid had been isolated, and a number of compounds related to phenol called cresols were also available from the distillation of crude creosote. These were the "therapeutic agents" in the Vapo-Cresolene lamp, advertised as being a boon to asthmatics who would be cured as they slept. But there was no cure. Asthma is not caused by germs. While the Vapo-Cresolene's dispensing of a supposedly healing vapor seems similar to my experience, clearly I wasn't inhaling cresols. So what was I breathing in?

A hunt through a pharmacology text quickly focused my attention on Dr. Schiffmann's Asthmador, a greenish powder. That meshed with my memory. Sold until the 1960s, Asthmador was to be burned much like incense and its fumes inhaled to relieve the symptoms of asthma. The active ingredient was atropine, a compound found in plants of the nightshade family, particularly *Datura stramonium* and *Atropa belladonna*. As early as 3400 B.C., Egyptian physicians recommended placing the dried leaves of such plants over heated bricks and inhaling the fumes to help with breathing problems. They were also aware that the juice of the nightshade plants would dilate the pupils when placed in the eyes, a phenomenon later capitalized on by Roman ladies who believed that enlarged pupils made them look more beautiful. Hence the name "belladonna," for "beautiful lady."

By 100 A.D., in India, special pipes were handcrafted for the burning of datura leaves. Since inhaling atropine vapors can produce hallucinations, it isn't clear whether the pipes were meant to ease breathing problems or to brighten lives. What we do know, is that in 1802, British physician James Anderson, who suffered from asthma, visited India and found relief by smoking datura. On his return to Britain, he related his experience to colleagues, one of whom, a Dr. Sims, published a report about the benefits of smoking datura in the *Edinburgh Medical and Surgical Journal*.

Before long, powders to be smoked in pipes, as well as cigarettes filled with datura or belladonna leaves, appeared on the market. Sounds like what I may have benefitted from. Turns out that atropine blocks receptors for the neurotransmitter acetylcholine, a process that dilates bronchial tubes. Such dilation relieves the symptoms of asthma. As with any chemical, dose matters, and numerous cases of overdose, as well as wilfull abuse by people seeking a high, have been reported. Dr. Sims himself apparently died from an overdose of belladonna, a plant that has justifiably earned the name "deadly nightshade."

Recently, the deadly nature of belladonna has been in the news again with the Food and Drug Administration (FDA) in the U.S. raising alarm about homeopathic teething pills. There is a suspicion that these may have caused seizures in babies and possibly even resulted in some deaths due to belladonna toxicity. Why belladonna should be associated with such products is bizarre to start with. According to homeopathic doctrine, if a substance is to relieve pain at a homeopathic dose, it should cause pain when used at a high dose. While atropine can cause many adverse symptoms, pain isn't one of them.

Furthermore, according to the tenets of homeopathy, dilutions are so extreme that none of the original substance remains in solution. So how can "nothing" cause harm? It seems some homeopathic companies are not very adept at making dilutions, and the effects on the babies were likely caused by an overdose of belladonna. In any case, the probability that homeopathic teething "remedies" will relieve babies' pain is about the same as the chance that the Vapo-Cresolene lamp can cure asthma.

BACTERIA ARE NOT ALWAYS BAD

One of the most significant scientific advances of the nineteenth century was the germ theory of disease. The twenty-first century may well be remembered for the germ theory of health.

The "germs" in this case are bacteria, specifically the ones with which we share our bodies. They live on our skin, in our mouth and nose, but it is in our colon that they really frolic. There, hundreds of species dine on the remnants of our meals and proceed to crank out compounds in their excreta that can enter our bloodstream. These may play a far more important role in determining our health than we ever envisaged.

In the womb, babies cavort in a sterile environment. Not a bacterium in sight. But as they emerge into the world, the first living species they encounter are the bacteria that inhabit the birth canal. These microbes are passed on from mother to baby, and together with others picked up from breast milk, food, water, pets, soil, and other people, eventually colonize the body, particularly the gut. Collectively referred to as the "microbiome," these bacteria multiply until their cells outnumber human cells by about ten to one. If we go by number of cells, we are only about 10 percent human and 90 percent bacteria.

An increasing number of studies now suggest that the microbiome may play a role in protection against various diseases.

For example, babies born by cesarean section, and therefore not exposed to the large variety of bacteria present in the birth canal, are at greater risk of becoming host to "bad" bacteria, which may predispose them to diseases such as celiac disease, type 1 diabetes, and perhaps even to obesity.

Exactly how this happens isn't totally clear, but the answer may lie in chemicals made by some bacteria that leak into the bloodstream. For example, mice with autism-like symptoms have a different mix of gut microbes than normal mice. Perhaps chemicals produced by some of these bacteria find their way to the brain. It is noteworthy that when these mice are treated with *Bacteroides fragilis*, a beneficial species that may crowd out "bad" bacteria, their symptoms improve. Maybe the treatment of autism in humans will come from altering the microbial mix in the gut.

There's yet another aspect to early exposure to microbes. We've grown up with the idea that dirt is bad. If you drop food on the floor, don't dare to pick it up and eat it. Sanitize your kitchen and bathroom with "germ killers." Filter your water. Purify your air. Sterilize baby bottles. Well, maybe all that attention to fastidious cleanliness isn't serving us so well. Maybe our immune system needs the exercise of dealing with microbes.

Proponents of the "hygiene hypothesis" maintain that if our immune system is deprived of the targets it has evolved to deal with, namely microbes, it turns its weapons on whatever target is available, even if this target is not dangerous. That target may be a protein in peanuts or an ingredient in a perfume. Immune reactions often involve inflammation as the body rushes white blood cells to the site of a perceived attack by an intruder. Sometimes inflammation can become chronic and may even be implicated in diabetes, heart disease, and stroke. Indeed, a study in the Philippines, where sanitation is not what we are used to here, showed that the more disease-causing microbes to which children are exposed, the lower their blood levels of C-reactive protein, a marker of inflammation, by the time they reach the age of twenty. For example, having spent time in a place with possible exposure to animal feces during childhood significantly reduces C-reactive protein levels in later life. Somehow it seems that early exposure to germs reduces the risk of chronic inflammation. And such exposure may even reduce the risk of cancer.

In fact, in the late 1800s, William Coley, a New York surgeon, made a remarkable observation. When he looked through surgical records, he found that at least as far as cancer surgery went, patients survived longer before the introduction of infection control methods. He was particularly taken by a patient who recovered from throat cancer after coming down with a bacterial infection. Coley began to wonder if somehow such an infection primed the body to attack tumors. To test his hypothesis, he injected cancer patients with *Streptococcus pyogenes*, the bacterium that had apparently saved the throat cancer patient. The injections worked, and Coley reported successes even when the bacteria were injected after tumors had metastasized. He noted that patients developed fevers in response to the bacterial injections, and the more dramatic the rise in temperature, the better the treatment appeared to work.

But then X-rays were discovered, and X-ray treatment of cancer patients showed so much promise that the Coley method was largely forgotten. It didn't help that many physicians looked upon Coley's ideas as quackery, being unable to resolve how an injection of a disease-causing bacterium could actually have a healing effect. As strange as it may seem, Coley may have been onto something. Dairy farmers, who shovel manure all the time and breathe aerosolized manure dust, have a five times lower rate of lung cancer than the general population. And it is specifically dairy farmers. Colleagues who work in the fields and orchards don't get this benefit. Furthermore, the greater the number of cows they work with, the greater the protection. Manure contains bacterial by-products called endotoxins that are used by the immune system to zero in on bacteria. Exposure to these apparently revs up the immune system and gets it ready to attack cancer cells, which are targeted because they produce endotoxin-like compounds.

This argument is supported by an examination of cancer rates among female cotton textile workers in China. Cotton dust contains lots of endotoxins, and it turns out that women with higher and longer endotoxins exposure have a lower incidence of lung, breast, liver, stomach, and pancreatic cancer. And then we have the interesting finding that children who are in daycare during their first few months of life are less likely to develop leukemia or Hodgkin's lymphoma as young adults. Studies have also shown that people who received a tuberculosis or smallpox vaccine as children have a lower risk of developing melanoma. Vaccines of course work by boosting the immune system. So the theory that seems to be developing is that cancer occurs when the immune system doesn't recognize cancer cells as dangerous because it has not been programmed properly in early life by exposure to a variety of microbes. An interesting theory, of course in need of further proof. But who would have ever thought that the answer to cancer may lie in shoveling sessions in a cow barn?

DON'T TAKE A DEEP BREATH

The crowd that gathered at St. James Hall in London on a November evening in 1884 was not your usual concert audience. Many of the seats were occupied by scientists who had been invited to witness the effects of an invention that was claimed to have the "most wonderful results upon the throat and lungs, and would also extend the range of the voice, making the notes full and rich. Dr. Carter Moffat, a professor of chemistry in Glasgow, was set to introduce his Ammoniaphone, a device he had been working on for years that was said to be a triumph of chemical science. Since childhood, Moffat had been interested in improving the speaking and singing voice since his own was weak, and as he described it, of "very poor quality and almost destitute of intonation." In order to improve it, he carried out experiments inhaling numerous gases and partook of various chemical solids and fluids. This had set the stage for his discovery.

The concert was organized by Miss Carlingford, a young vocalist who had garnered fame with her performances in Gilbert and Sullivan operettas. She now wished to publicly recognize the benefit she and many of her professional friends had derived from the use of the instrument.

During a break in the performance, Dr. Carter stepped to the podium and delivered a lecture on how he had developed the Ammoniaphone after wondering why Italian singers had such golden tones. On a professional visit to Italy, he claimed to have noted that the air there had a different quality, prompting him to examine it in various locations around the country. He concluded that Italian air had unusually high levels of hydrogen peroxide and ammonia, probably the result of volcanic activity. A curiosity, since at that time there was no way that the trace quantities of these chemicals in air could have been identified. Moffat never did describe his methods, but before long he had designed his Ammoniaphone, a flute-like tube containing an absorbent material saturated with hydrogen peroxide at one end and ammonia at the other. There was a mouthpiece at the center through which a current of air could be drawn into the lungs. Moffat claimed that after a few whiffs squeaky voices would be converted into robust tenors. Once depleted of chemicals, the instrument could be sent to the manufacturer to be recharged.

After Dr. Carter's presentation, volunteers were invited on stage and asked to speak in their natural voice before inhaling the "Artificial Italianized Air." The inventor called attention to the fact that after using the Ammoniaphone the voices were louder and had an improved tone. Not everyone, though, was convinced. One newspaper correspondent noted no obvious change and remarked "that if the inhalation of free ammonia and peroxide of hydrogen is so good for the voice, it seemed scarcely necessary to enclose these ingredients in an expensive flute-like case to test their powers, and the fact of doing so and calling the vapor they give off 'artificial Italian air' savours to me of quackery."

Moffat licensed the production of the Ammoniaphone to the Medical Battery Company that then geared up the advertising, recommending the gadget not only for singers but for public speakers, parliamentary men, schoolmasters, and clergymen as well. The company even commissioned an Ammoniaphone song that told of the plight of a young man who wanted to propose to his sweetheart but had lost his voice:

"Ah! Well for him and for the fair, / He'd heard that pure Italian air / Might be inhaled imparting tone, / Through Moffat's famed 'Ammoniaphone.'"

The device claimed to do more than just "cultivate voice by chemical means." Throat and chest diseases such as asthma, bronchitis, cough, deafness from colds, and even sleeplessness would succumb to its powers. Interestingly, another contraption sold by the same company had a similar promise. That was

"Dr. Carter Moffat's Cool Featherweight Electric Body Belt." "Stop taking poisonous drugs," the ads urged, "instead wear the electric belt and say goodbye to indigestion, liver torpidity, internal weakness, gout, sciatica, sleeplessness, melancholia, palpitation, and other drug baffling ailments. Should be worn by all because its electricity is absorbed by the system. No vinegar or other acids need be used. It removes morbid and impure matters from the blood. Great for nervous depression and brain overwork."

Certainly people who bought into this daft scheme did not suffer from an overworked brain. How Dr. Carter, a respected scientist, got roped into this scheme isn't clear, but perhaps by licensing his Ammoniaphone to the Medical Battery Company he had signed a deal with the devil. In 1893, the company was indicted for fraud and put out of business and with that the Ammoniaphone faded into obscurity, relegated to museum showcases. Visitors to London's Science Museum can still see one, although most just walk by, oblivious of how this peculiar tube once captivated a nation.

There's no question that many singers sang the praises of the Ammoniaphone. Could it really have improved their warbling? Certainly, inhaling a gas can change the sound that is produced. It isn't because the vocal chords vibrate at a different frequency; their vibration is actually independent of the gas that surrounds them. But the speed at which sound travels depends on the density of the gas. Helium is less dense than nitrogen, the major component of air, and that is responsible for the classic Donald Duck effect. It is possible that inhaling a mix of ammonia and vapors of hydrogen peroxide would temporarily change the timbre of the voice and that is what fans of the Ammoniaphone experienced. That theory would require testing, but as much as I like science, I'll leave the exploration of inhaling "Italianized air," which according to Dr. Carter was "the most beneficent discovery the world has ever known," to someone else.

PHOSPHIDES AND BEDBUGS

Dazzled by its greenish glow and aware of its original isolation from urine, some called phosphorous "the vital flame of life." And within a few years of its 1669 discovery by the German alchemist Hennig Brandt, the radiant element took on the role of protecting that flame from the deluge of disease. Dr. John Ashburton Thompson's 1874 opus, "Free Phosphorus in Medicine," described the treatment of colic, gout, tuberculosis, and even "mental instability." The element was even said to produce "venereal excitation," the common euphemism of the day for "sexual stimulation." French apothecary Alphonse Leroy apparently put that notion to a test and reported a positive effect in listless ducks. Thompson tried small doses of phosphorus himself but failed to produce a duck-like response. Aware that larger doses of the element caused nausea, bad breath, and loosened teeth, he wondered if compounds of phosphorus would be more suited to his needs.

Small packets of powdered zinc phosphide had been available since 1867 as a remedy for whatever ailed the patient. Dr. Thompson gave this a shot but instead of venereal ardor, only a smelly burp emerged. That's because zinc phosphide reacts with stomach acid to produce phosphine, a gas with a decidedly garlicky odor. Luckily for Thompson, he swallowed only a few milligrams of the phosphide and thus managed to avoid a permanent squashing of all passions. Phosphine is potentially lethal! However, that is just what makes the gas useful as a fumigant. By the 1930s, phosphine released from aluminum or magnesium phosphide tablets when these react with moisture in the air was ridding grain elevators and transport containers of insects. And when it came to rodents, pellets of aluminum phosphide combined with bait made for a last supper.

In 2000, workers in a German office building almost met the fate of insects. They noticed a strong smell of garlic and soon began to feel nauseous, headachy, and complained of sore throats. When a plant in the office suddenly lost all its leaves, it was time to call the fire department. The firemen recognized the garlic smell and suspected phosphine had been released. Seven workers were rushed to hospital and the street was quickly evacuated.

The source of the gas turned out to be the tobacco store next door. Its owner had been importing cigars from the Dominican Republic and discovered they were infested with the tobacco fly. His Dominican contact supplied him with aluminum phosphide pellets, which he spread on the floor one Friday, hoping to eliminate the flies over the weekend. He almost ended up eliminating the occupants of the adjacent building. Luckily the firemen were well-trained to recognize the smell of phosphine!

The outcome of a similar situation in Jerusalem in 2014 was not so lucky. Two young children died and two others ended up in hospital in critical condition due to a fumigation process gone terribly astray. A problem with insects or rodents in an apartment prompted a call to an exterminator who chose to use aluminum phosphide to generate phosphine gas. Bad idea. Apparently the room in the apartment that was fumigated was sealed with plastic sheeting, but the family was not told to leave. The seal turned out to be inadequate. And phosphine seeped out. Being heavier than air, it affected children the most readily.

The sick children were taken to a clinic, where the parents informed the doctor that they thought the problem was some bad food. There was no mention of any insecticide being used, so there was no way for a physician to consider the possibility of this type of poisoning. After being examined at the clinic, the children appeared to be fine and were sent home. Eventually emergency personnel were called in only to find one child dead and the other occupants of the apartment in dire straits. Seems there was further exposure to leaking phosphine, for which there is no antidote. The exterminator was arrested, accused of negligence.

A similar tragedy struck recently in Fort McMurray, Alberta, where a family's battle with bedbugs resulted in the death of an infant. Bedbugs may suck your blood at night and leave annoying bite marks, but they won't kill you. But trying to kill them with phosphides can. These chemicals are not available for home use in Canada but can be readily purchased in some other countries. It appears the phosphide used in this tragic case was brought back from Pakistan. Sadly, the phosphine it released ended up extinguishing the vital flame of life.

METHYLENE BLUE MAGIC

In 1976, the catchy tune "(Shake, Shake, Shake) Shake Your Booty," by KC and the Sunshine Band quickly rose to number one on the charts. The song stirred up a little controversy, with some complaining of a sexual connotation, but for me the song will forever be linked with methylthioninium chloride, a chemical you may be more familiar with by its common name, methylene blue.

The song, which has the distinction of being the only number-one song with a word repeated more than three times in its title, came out the same year that colleague Ariel Fenster and I conceived "The Magic of Chemistry," a stage show that blended chemical demonstrations, slides, magic, and music with the aim of bringing chemistry to life in an entertaining way. One of the demonstrations we featured was a classic known as the blue bottle experiment. This features a colorless solution in a flask that turns a dark blue upon shaking. When the shaking stops, the color slowly disappears. It can be restored by shaking again. The chemistry involved is fascinating.

Methylene blue is a crystalline substance that yields a blue solution when dissolved in water but the solution turns colorless upon the addition of glucose. On shaking, the methylene blue reacts with oxygen, restoring the blue color. Then it turns colorless again as it reacts with glucose. The cycle can be repeated until all the glucose is used up. I was looking for some music to accompany the shaking when almost miraculously "Shake Your Booty" hit the airwaves. It was perfect! We've been shaking the booty out of that methylene blue ever since.

But this dye can do a lot more than shake up an audience at a chemical magic show. Our story takes us back to the Victorian era, basically the second half of the nineteenth century, one of the most colorful eras in history. Literally. It was marked by the introduction of a host of synthetic dyes that to a large extent replaced the natural dyes derived from plants. The dye revolution was sparked by the accidental synthesis of a brand new color, mauve, by young William Henry Perkin, who was actually trying to make the antimalarial drug quinine from chemicals found in coal tar. The attempt to make quinine was futile, but Perkin recognized the value of mauve and with his father's help set up a successful dye factory.

Heinrich Caro was a chemist in Germany working with natural colorants when he was sent by his company to England to learn about synthetic dye production. Instead of returning, he found employment with a Manchester chemical company but was eventually lured back to Germany, becoming the first head of research at Badische Anilin- und Soda-Fabrik, known by the abbreviation BASF. It was here that Caro managed to synthesize methylene blue, a novel dye for cotton.

The new synthetic dyes found a use beyond coloring fabrics. Medical researchers discovered that they were also capable of selectively staining different kinds of cells and microbes for easy visualization under a microscope. In 1887, Polish pathologist Czesław Chęciński found that methylene blue stained the malaria-causing parasites *Plasmodium malariae* and *Plasmodium falciparum*. Subsequently, German physician and scientist Paul Ehrlich noted that methylene blue would not only stain the parasite, it was capable of killing it.

By 1891, Ehrlich was using the dye to treat victims of malaria, making methylene blue the first-ever synthetic drug used in medicine. This was a big advance given that quinine, the classic antimalarial, had to be isolated from the South American cinchona tree while methylene blue could be produced on a large scale. But methylene blue wasn't quite as effective as quinine, a problem tackled by Ehrlich's student Wilhelm Rohl, who worked at Bayer. Rohl did what chemists normally do when trying to improve a drug, namely, slightly alter its molecular structure. This research eventually gave rise to quinacrine, an effective antimalarial. Then in 1934, Hans Andersag at Bayer modified quinacrine further and synthesized chloroquine, which would eventually become the standard treatment for malaria and is used to this day. Unlike methylene blue or quinacrine, chloroquine didn't color the skin or the eyes blue.

Methemoglobinemia is a condition in which hemoglobin, the oxygen-carrying compound in the blood, is converted to methemoglobin, an altered form that is less capable of delivering oxygen to tissues. While there are hereditary forms of this condition, it is more commonly caused by exposure to chemicals that include certain antibiotics, anesthetics, nitrates, or aniline dyes. Interestingly, a treatment for this ailment is intravenous methylene blue.

Methylene blue may yet have another application. It is taken up in the nervous system by aggregations of tau protein, one of the hallmarks of Alzheimer's disease. Claude Wischik, while at Cambridge University, discovered that methylene blue not only stained the tau proteins but untangled them. He eventually founded TauRx, a Scottish pharmaceutical company that has recently released the results of a placebo-controlled trial using a slightly altered version of methylene blue, code-named LMTX, on Alzheimer's patients. After fifteen months, subjects taking LMTX deteriorated significantly more slowly than those on placebo. In a few cases there was even improvement. Significantly, MRI scans documented that LMTX was able to slow the brain atrophy normally associated with Alzheimer's disease. Interestingly, the benefits of LMTX were not seen in patients also taking standard Alzheimer's drugs, possibly because these cause rapid clearance of LMTX from the bloodstream.

On this side of the ocean, researchers at University of Texas found memory improvement in a small placebo-controlled study of thirteen healthy adults and also showed, via functional magnetic resonance imaging, that methylene blue can cause increased activity in areas of the brain related to memory. If this synthetic compound pans out as a memory enhancer and possible Alzheimer's treatment, it will add another vivid chapter to methylene blue's colorful history.

ARSENIC ARCHIVES

During a recent talk on the relation between the body and the mind, I mentioned the newest anxiety-relieving craze, coloring books. Aimed at adults, these feature intricate patterns that provide quite a challenge for staying inside the lines. The contention is that focusing on the special patterns distracts the mind from anxiety and stress. Evidence is sketchy, but millions of coloring books are flying off the shelves, topping bestseller lists. That in itself says something about our society.

After my talk, I was approached by a lady who claimed she had something better than coloring books to relieve anxiety and slipped a vial full of pills into my hand. She didn't seem like a clandestine drug pusher, so I thought I would look down and find some pills of lorezepam or maybe St. John's Wort. Such was not the case. The label on the vial read "Arsenicum album 30C."

No, she was not trying to poison me. These were homeopathic arsenic pills. Recall that homeopathic theory relies on the curious notion that a substance that in large doses causes certain symptoms can, in homeopathic potency, repel the same symptoms. Since arsenic poisoning is associated with anxiety and restlessness, a person suffering such symptoms should find relief in a homeopathic dose of arsenic. In the bizarre world of homeopathy, potency increases with greater dilution, and a dose of 30C is said to be extremely potent. Such a pill is made by sequentially diluting a solution of arsenic a hundred fold thirty times and then impregnating a sugar pill with a drop of the final solution. At a dilution of 30C, not only is there no trace of arsenic left, there isn't even a water molecule that has ever encountered any of the original arsenic.

Homeopathy is a scientifically bankrupt practice that was invented over two hundred years ago by German physician

Samuel Hahnemann, who was disenchanted with bloodletting and purging, common medical procedures at the time. He was a good man who searched for kinder and gentler treatments and homeopathy fit that rubric. Since knowledge of molecules was almost nonexistent at the time, Hahnemann could not have realized that his diluted solutions contained nothing. Actually, the truth is that they did contain something. A hefty dose of placebo!

Now here is the kicker to this story. Hahnemann was quite accomplished in chemistry and actually developed the first chemical test for arsenic. In 1787, he found that arsenic in an unknown sample was converted to an insoluble yellow precipitate of arsenic trisulfide on treatment with hydrogen sulfide gas. When in 1832 John Bodle in England was accused of poisoning his grandfather by putting arsenic in his coffee, John Marsh, a chemist at the Royal Arsenal, was asked to test a sample of the coffee. While he was able to detect arsenic in the coffee using Hahnemann's test, the experiment could not be reproduced to the satisfaction of the jury and Bodle was acquitted. Knowing that he could not be tried for the same crime again, he later admitted to killing his grandfather.

The confession infuriated Marsh and motivated him to develop a better test for arsenic. By 1836, he had discovered that treating a sample of body fluid or tissue with zinc and an acid converted any arsenic to arsine gas, AsH_3, which could then be passed through a flame to yield metallic arsenic and water. The arsenic would then form a silvery-black deposit on a cold ceramic bowl held in the jet of the flame and the amount of arsenic in the original sample could be determined by comparing the intensity of the deposit with a sample produced with known amounts of arsenic.

The Marsh test received a great deal of publicity in 1840 when Marie Lafarge in France was accused of murdering her husband by putting arsenic into his food. Marie was known to have bought arsenic from a local chemist. She claimed it was to kill rats that had infested the house. A maid swore that she had seen her mistress pour a white powder into her husband's drink and Marie had also sent a cake to her husband who was traveling on business just prior to his becoming ill. The dead husband's family suspected that Marie had poisoned him and somehow got hold of remnants of food to which she had supposedly added arsenic. The Marsh test revealed the presence of arsenic in the food and in a sample of eggnog, but when the victim's body was exhumed the investigating chemist was unable to detect arsenic.

To help prove Marie's innocence by corroborating the results of the investigation of the exhumed body, the defense enlisted Mathieu Orfila, a chemist acknowledged to be an authority on the Marsh test. Much to the defense's chagrin, Orfila showed that the test had been carried out incorrectly and used the Marsh test to conclusively prove the presence of arsenic in Lafarge's exhumed body. Marie was found guilty and sentenced to life in prison. The controversial case captured the imagination of the public and was closely followed through newspaper accounts, making Marie Lefarge into a celebrity. It would also go down in the annals of history as the first case in which a conviction was secured based on direct forensic toxicological evidence. Because of Mathieu Orfila's role in the case, he is often deemed to be the "founder of the science of toxicology." The Marsh test became the subject of everyday conversations and even became a popular demonstration at fairgrounds and in public lectures. This had an interesting spin-off: poisonings by arsenic decreased significantly since the existence of a proven, reliable test served as a deterrent.

As far as claims about relieving anxiety with homeopathic

arsenic go, well, they cause me anxiety. I think I'll flush those homeopathic tablets down the drain (no worry about arsenic pollution here) and buy a coloring book.

SOCKS, WALLPAPER, AND ARSENIC

Brightly colored socks are "in." I'm all for this fashion trend since I've always had a soft spot for the chemistry of dyes. I have no concern about wearing vibrant socks, but that would not have been the case back in the 1860s when multicolored socks first appeared on the scene thanks to the new synthetic dyes pioneered by William Henry Perkin's accidental discovery of mauve. Within a few years of that epic moment in 1856, a whole range of dyes manufactured from coal tar hit the fashion scene. And before long, reports of toxic reactions hit the media.

An 1868 edition of *The Lancet*, the leading medical journal of the time, featured a report by a physician who had noted that "the dye used in some of the gorgeous socks and other underclothing displayed in the windows of some of the metropolitan hosiers exercises a very deleterious influence upon the skins of the wearers, producing irritation and an eruption upon the skin." The article also mentions the case of a ballerina performing at the famous Drury Lane theater who developed an "anomalous eruption" on one foot and not the other. The mystery was solved by the revelation that the performance required her to wear different-colored socks. It was the foot adorned with a brilliant red sock that was affected! Dr. Webber's report in *The Lancet* was picked up by newspapers, and the resulting public outcry caused the manufacturer to stop selling the multicolored socks and revert to using natural Brazilwood and logwood dyes.

Some of the British socks, though, had been exported to France, where they came to the attention of a Dr. Tardieu. It seems a number of his patients had developed a rash on their feet after wearing the imported socks with red stripes. Tardieu began an investigation into the "poisoned socks" by enlisting the help of Monsieur Roussin, a chemist who extracted the red dye from the socks with alcohol and identified it as coralline. This synthetic dye was produced by reacting oxalic acid with phenol, a chemical isolated from coal tar. To investigate its toxicity, Tardieu injected the extracted coralline under the skin of a dog, a rabbit, and a frog. All three animals died. He published his findings in the *Gazette des Hôpitaux*, causing coralline to quickly fall out of favor with manufacturers who did not want to be associated with toxic socks.

It wasn't long before Tardieu's allegation of the poisonous nature of coralline was challenged. Dr. Landrin, a veterinary surgeon, reported that he had injected dogs and cats with pure coralline without any ill effects, an observation confirmed by a second veterinarian. How could coralline be poisonous in one case and not in another?

That conundrum was solved by a singular incident, detailed in 1874 in the *British Medical Journal.* French physician Dr. Bijon suffered from "prickings of the eyelids, with itching and burning sensations" that were exacerbated in a particular room in his house. The room had a different kind of wallpaper from the others, arousing Bijon's suspicion since he was aware of accounts that had linked a dye known as Scheele's green with all sorts of health problems. In 1775, Carl Wilhelm Scheele had combined arsenic oxide, sodium carbonate, and copper sulfate to produce copper arsenite, a stunning green substance that was widely used to dye fabrics and wallpaper. Arsenic-bearing dust from dyed garments and wallpaper tainted rooms, and the situation was further worsened by the moist air in Victorian homes that was conducive to the buildup of mold. Various molds can feed on Scheele's green and release volatile and poisonous arsine gas in the process.

Although Bijon did not have green wallpaper, he still wondered about the possible presence of arsenic and had his wallpaper analyzed. Two chemists, using the classic Marsh test, confirmed that arsenic was indeed present along with coralline in the wallpaper's red pattern. But coralline does not contain arsenic! The eventual conclusion was that an arsenic compound had been used as a mordant. Deriving from the French verb "to bite," mordants are inorganic compounds that bind a dye to a fabric. Dr. Tardieu had been unaware of this, and when he injected his animals with the sock extract, he was actually poisoning them with arsenic, not coralline.

There is no question that arsenic poisoning affected many people in the eighteenth and nineteenth centuries, but the case that still captivates the public's imagination the most is one where arsenic poisoning is suspected but has never been proven. That is the case of Napoleon, who according to some experts died of a stomach bleed, a consequence of stomach cancer caused by arsenic. It is a fact that during his exile on Saint Helena, the emperor lived in a room with bright green wallpaper, and that the climate on Saint Helena is damp, making mold growth a decided possibility. It is also true that arsenic can cause cancer.

Some of the fourteen samples clipped from Napoleon's hair the day he died did show contamination with arsenic, but since not all of the samples had arsenic, critics of the arsenic poisoning theory suggest that contamination came from later efforts to preserve the samples. Furthermore, Napoleon did not have skin lesions and showed no sign of nerve damage or weight loss, all typical symptoms of arsenic toxicity. In fact, when he died, he was considerably overweight, no longer the "Little Emperor." Because of his stomach problems, Napoleon also took large doses of calomel (mercurous chloride), which could have caused the stomach bleed that killed him.

Neither Scheele's green nor arsenic mordants are used these days, and coralline, probably unjustly accused of causing rashes, has also been replaced by a bevy of modern nonirritating synthetics. So there is no need to be concerned of any toxicity when telling an annoying person to put a sock in it.

A RAT POISON THAT CAN CURE

One of the most often told chemical stories, at least by me, is the accidental discovery of the first synthetic dye in 1856 by William Henry Perkin at the ripe old age of eighteen as noted earlier. During a futile attempt to synthesize quinine from compounds found in coal tar, young Perkin accidentally produced a substance that had a stunning mauve color. Up to that time, dyes for fabrics were extracted from a natural source, but this changed with the discovery of mauve. Before long a number of other "coal tar" dyes appeared on the marketplace, many produced by Perkin, who with financial help from his father and brother opened a dye factory. By the time he reached his thirties, Perkin was wealthy enough to retire from the dye business in order to devote all his time to chemical research. His main interest, as one would expect, focused on synthetic chemistry, that is using chemical reactions to make novel compounds.

One of Perkin's targets was coumarin, much desired by the perfume industry. A synthetic version would make perfume production much easier than having to extract the compound from a natural source such as the tonka bean. This is actually the seed of the tonka fruit that grows on a tree native to Central and South America. When dried, it has a pleasant vanilla-like odor that can be traced to its content of coumarin, first isolated from the bean in 1820. The term "coumarin" comes from "cumaru," the name of the tonka tree in a native tongue. There have been cases of tonka beans being used fraudulently to substitute for the more expensive vanilla.

Coumarin has liver toxicity in rats but it is unlikely that humans would ever consume enough coumarin to cause a problem. In fact in Europe shavings of the tonka bean are sometimes used to flavor food, a practice not permitted in Canada or the U.S. The major use of coumarin is in the manufacture of perfumes but for this it is not extracted from a natural source; it is produced synthetically. Indeed coumarin was the first synthetic substance ever used in perfume production. Credit for this goes to Perkin, who in 1868 managed to make coumarin from compounds isolated from coal tar. This hinged on a reaction he invented that was later christened the "Perkin reaction," a staple in organic chemistry courses.

Synthetic coumarin first appeared on the market in 1882 in Fougere-Royal, the highly successful perfume that became the prototype for a host of "Fougères," linked by the inclusion of coumarin. The compound didn't get much attention outside the perfume industry until the 1920s when cattle in Canada and the northern U.S. began to die from some mysterious condition that caused them to bleed to death internally. The connection turned out to be the consumption of moldy silage made from sweet clover. Before long, researchers using rabbits showed that moldy hay, but not regular hay, had an anticoagulant effect and by 1940 Professor Karl Paul Link at the University of Wisconsin managed to isolate the substance that was responsible. Link's interest was originally piqued when a farmer showed up in his lab with a milk can full of uncoagulated blood and a cow that had died after eating moldy sweet clover.

Suspicion was that coumarin was the culprit, but assays on rabbit blood showed it had no anticoagulant effect. Link then went on to show that chemicals in mold can convert coumarin into dicoumarol, a compound that interferes with the action of vitamin K, the substance that prevents excessive bleeding by coagulating blood when necessary. This work came to the attention of the pharmaceutical industry, which was searching for alternatives to heparin, the only anticoagulant available at the time. Dicoumarol was introduced into medical practice in 1941 to treat patients who had suffered strokes, had heart valve problems, or had irregular heartbeats. But some companies also promoted the substance as a rat poison. This concerned Professor Link because he thought it could doom the use of dicoumarol as a drug since patients might be reluctant to take a "rat poison."

During his years of research attempting to identify the anticoagulant present in moldy sweet clover, Link had synthesized a number of coumarin derivatives. He now reexamined these to see if any could be used as a rat poison, hoping to steer manufacturers away from using dicoumarol for this purpose. Compound #42 seemed to fit the bill and Link named it "warfarin" from "coumarin" and the acronym WARF, for the Wisconsin Alumni Research Foundation, which had supported his research.

Warfarin was first registered for use as a rodenticide in the U.S. in 1948 and caught pharmaceutical chemists' attention in 1951 when an American soldier tried to commit suicide with warfarin and was saved when he was given vitamin K. Dicoumarol had no antidote and that had been a problem in its clinical use. Now the success with an antidote suggested that warfarin could be used as a drug in humans when blood clot formation needed to be prevented. It did indeed prove to be superior to dicoumarol, and ironically, given Link's concern about using a rat poison as a drug, was approved for medical use in humans in 1954. Given the trade name Coumadin, the drug garnered extensive publicity when it was prescribed for President Eisenhower after his heart attack in 1955. It is still the most widely used anticoagulant in the world. And it is still used as a rat poison. A rat poison that saves human lives. Another classic example of how the same chemical can be used to kill or cure. The choice of how to use it is ours.

THE NUREMBERG CHRONICLE

It isn't often that you get a present that is over five hundred years old. I'm not talking about some fossil or gemstone or ancient Roman coin. What I was fortunate to receive was an original page from a book commonly known as the *Nuremberg Chronicle*. Compiled by the German physician Hartmann Schedel and printed in 1493, the book chronicles the history of the world since "the first day of creation" to Schedel's time. The text leans heavily on the bible but also focuses on the history of a number of cities and gives detailed accounts of the lives of saints, prophets, kings, popes, and mythical heroes. There are also descriptions of freaks of nature and postulations about the coming of the Antichrist and judgment day. But the true fame of the volume rests on the integration of over 1,800 illustrations with the text. For the first time people were able to see history unfold with pictures.

The page I now possess describes with appropriate illustrations the instructions as given by God to Moses for the construction of an incense-burning altar and for building the Tabernacle, a portable shrine designed to house the Ark of the Covenant as the Israelites wandered through the wilderness.

As I inspected this amazing relic, I was struck by the remarkable clarity of the text and illustrations. Although the ink had been applied to the paper over five hundred years ago, it showed no signs of age. That, I thought, presented an opportunity to shine the spotlight on ink, a substance that, let us just say, doesn't get enough ink. When it comes to inventions that have made the greatest impact on human history, ink surely ranks up there with fire, the wheel, gunpowder, and electricity. We wouldn't have much recorded history if we had to rely on carving words into clay or stone.

Basically, there are three types of ink. The oldest variety is based on a water-insoluble substance, called a pigment, most commonly finely powdered carbon, or "lampblack," collected as soot from burning oil. This is mixed with water and gelatin from animal hides or with a gum, a water soluble carbohydrate extract from a plant or tree. Gum arabic from acacia trees is a typical example. These sticky substances ensure that the pigment will adhere to the surface once the water evaporates. By 2500 B.C., both the ancient Chinese and Egyptians were using such inks, later called India ink, as the ingredients were often sourced from India. Another ancient pigment, mercury sulfide, or cinnabar, was detected in the red ink on the famous Dead Sea Scrolls, which also featured iron gall ink along with the more common India ink for the black script.

This second kind of ink, iron gall, was first described in detail by Pliny the Elder in the first century, and by the fourth century it was rivaling lampblack in popularity. It was made by combining naturally occurring ferrous sulfate ("green vitriol") with tannic acid extracted from oak galls, growths on oak trees in response to chemicals secreted by wasp larvae. Ferrous sulfate and tannic acid form a dark soluble complex that becomes insoluble as the ferrous ions convert to ferric ions by giving up electrons to oxygen in the air. As with carbon inks, gum Arabic was added to help the ink remain in place on the paper, or parchment, the processed sheepskin that scribes preferred.

The third type of ink is based on dyes dissolved in a solvent. The earliest ones relied on natural inks secreted by squid, cuttlefish, or octopi or on extracts from berries or indigo plants. But practically, dyes became important only after William Henry Perkin's 1856 discovery of mauve, the first synthetically produced colorant. Dye-based inks were ideal for the emerging fountain pen industry since they didn't clog up the pen tips the way that pigments did. Ballpoint pen ink also uses dyes dissolved in a solvent such as propylene glycol, but the key here is the adjustment of the thickness of the solution to allow for a smooth flow. This is accomplished by a slew of additives with nitrocellulose-based polymers playing a role. Without a doubt, though, since the middle of the fifteenth century, the most important inks have been those used for printing.

While ink was always the key to disseminating information, its importance took a giant leap with Gutenberg's invention of the printing press around 1470. Water-based iron gall or India inks smudged when used with Gutenberg's moveable type, so he turned to ink made with soot, turpentine, and walnut oil. Modern analysis of the ink used for Gutenberg's original bible found that he also added oxides of copper and lead to increase the blackness. Today's printing ink is not much different, although the solvents used tend to be methanol, acetone, propylene glycol, hexane, xylene, or toluene, and flax seed oil or synthetic alkyd resins have replaced walnut oil.

The ingenuity that went into developing the ink used to print those early volumes such as the *Nuremberg Chronicle* is admirable. The content of the *Chronicle* is less so. While discussions of the works of ancient physicians such as Galen, Hippocrates, and Avicenna are more or less accurate, as are descriptions of natural phenomena such as the falling of a meteorite on the Alsatian town of Ensisheim in 1492, there are also mythical events that are presented as factual. A famous woodcut of "The Burning of the Jews" is repeated several times in the *Chronicle* with a different reason given each time for why it was done. The most disturbing description is coupled with an illustration of the "Murder of a Christian Child," depicting stereotyped Jews draining the child's blood for use in Passover rituals. Of course, the "burning" is true enough — such atrocities have tainted history — but portraying the mythical killing of the child as being factual, in what at that time was the most authoritative book on world history, amounted to laying a cornerstone for future prejudice. Unfortunately, such prejudice is as indelible as the ink used to produce the *Nuremberg Chronicle*.

Today, a deluge of information is readily available in an inkless fashion online. But if anything, the problem of separating fact from fiction has become even more acute than when Hartmann Schedel compiled his remarkable work in 1493.

A RABBIT OUT OF A HAT

Pulling a rabbit out of a hat, well that's sort of old hat. But what about pulling a rabbit out of a woman's privates? That is exactly what happened in 1726 when Mary Toft, an illiterate servant, seemed to give birth to a litter of rabbits and assorted other animal parts. Incredibly, some physicians were taken in by the hoax that captured the imagination of England as well as that of King George I.

The remarkable story began when Dr. John Howard was called to Mary's home to assist in her labor, but instead of a baby, she delivered what looked like animal parts. That didn't end the pregnancy though, and Howard was repeatedly called back over the next month to deliver first a rabbit's head, then the legs of a cat, and finally a litter of nine dead baby rabbits. Howard didn't know what to make of this bizarre phenomenon and sought help from other doctors. When the King heard about the story, he immediately dispatched surgeon Nathaniel St. André to look into the matter. Much to his surprise, St. André also witnessed Mary giving birth to several dead rabbits and became convinced that some sort of supernatural event was occurring. King George thought that further investigation was warranted and sent German physician Cyriacus Ahlers to investigate. He too witnessed several rabbit births but smelled a rat. One of the rabbits still had dung pellets inside that contained corn and hay. Since Mary's uterus was unlikely to produce such crops, Ahlers reported to the King that he suspected a hoax.

By this time the story had become a national sensation, prompted by St. André's publication of *A Short Narrative of an Extraordinary Delivery of Rabbets* and Mary's curious explanation for her strange prodigies. She claimed to have been startled by a rabbit while working in a field, which caused her to have a strong desire for rabbit meat, but being very poor, she was unable to satisfy the cravings. People bought the story because at the time there was a belief that emotional disturbances during pregnancy could influence the developing fetus.

"Maternal impression" still held sway into the nineteenth century, exemplified by the case of Joseph Merrick, the "Elephant Man" who claimed that his deformity was caused by his mother being frightened by an elephant during pregnancy. It isn't totally clear what Merrick actually suffered from, but the theory is that he was the victim of two rare conditions, neurofibromatosis and Proteus syndrome.

In order to observe Mary more closely, St. André arranged for her to be taken to London, where other doctors could be privy to the remarkable phenomenon. Alas, no more rabbits were produced. Mary's story began to unravel when a porter was caught trying to sneak a rabbit into her room, explaining that he had been hired by Mary's sister-in-law. That prompted an official investigation, but Mary admitted nothing. Only when she was threatened with surgery to explore her insides did she finally confess that she had inserted the dead rabbits into her birth canal manually and managed to squeeze them out, making it appear as if she were giving birth. Why did she do it? The age-old answer, money. In the eighteenth century, it was common for people to pay to see human curiosities, and what could be more curious than a woman who had given birth to rabbits?

Mary spent a few months in jail and then returned to relative obscurity. St. André attempted to vindicate himself by taking out an ad in a newspaper apologizing for his "mistakes" and expressing hope that "people would be able to separate the innocent from those who have been guilty actors of this fraud." But the public was not forgiving. St. André's patients deserted him, and he eventually died a poor man.

Cartoonists had a field day with drawings of Mary spewing out rabbits surrounded by caricatures of gullible physicians who had swallowed Mary Toft's story. And that really is what makes this case so fascinating. Although medical education was still rather primitive at the time, there was certainly enough known about anatomy and reproduction to have dismissed Mary's rabbit births as claptrap. But it seems education wasn't then, and isn't now, a vaccination against folly. There are physicians today who advocate against vaccination and many who buy into homeopathy, coffee enemas, antineoplastons, alkaline water, energy healing, and other forms of woo that make no more sense than a woman giving birth to rabbits.

TAKING PULSE

Avicenna, the famed Persian scholar of the Middle Ages (981–1037), was a true polymath, with his writings encompassing astronomy, alchemy, mathematics, physics, geology, psychology, and medicine. He devised a system for testing medicines, stating that "the time of action of a drug must be observed, so that essence and accident are not confused." That was an early formulation of "association is not the same as causation," a dictum we endeavour to drill into students' heads today.

Avicenna also emphasized that experimentation must be done with the human body, for "testing a drug on a lion or a horse might not prove anything about its effect on man," a problem with which we still struggle today. His method of diagnosing illness often involved taking the pulse, noting that this could be affected not only by physical ailments but also by the patient's mental state. On dealing with a melancholy gentleman, Avicenna diagnosed "love sickness" by noting a flutter in the man's pulse at the mention of a particular address. He would be cured, the sage suggested, by the face of the girl who lived there.

As early as the seventh century B.C., physicians in India, China, and Egypt measured heart rate by means of taking the pulse and related its frequency, strength, and regularity to the patient's state of health. In the fourth century B.C., the Greek physician Herophilus used a portable water clock to take a pulse measurement, and a hundred years later Galen, the most famous of the Roman physicians, described twenty-seven characteristics for a single beat of pulse in his treatise entitled *De Pulsuum Differentiis*.

Galen was a proponent of the humoral theory, according to which four liquids, namely blood, phlegm, black bile, and yellow bile, flowed through the body and determined well-being. A balance between these humors was essential to retaining a healthy body and mind, and Galen believed that imbalances could be noted by changes in the pulse. Treatments to restore a healthy pulse included modifying dietary habits, increasing exercise, and using herbal medicines along with more invasive interventions such as laxatives and emetics to purge excess humors. Bloodletting was also practiced based on the belief that some forms of illness were caused by excess blood. Humoral theory persisted in various forms until the nineteenth century.

Although taking the pulse had become a common procedure by physicians, there was no accurate way of measuring its frequency until Dr. John Floyer asked horologist Samuel Watson to make a watch with a hand that would rotate about the face once a minute. This then allowed the beat of the pulse to be accurately measured as described in Floyer's 1707 epic, *The Physician's Pulse Watch*, in which he related the pulse to respiration rate, temperature, barometric readings, age, gender, and even seasons of the year.

If Floyer detected a problem with the pulse, he would often prescribe cold bathing, a treatment that had religious overtones. Just as the Great Flood had purified the earth, a cold bath would purify the body. Floyer also was a proponent of the healing abilities of the "royal touch" and suggested that one of his young patients, Samuel Johnson, who would later become the famous lexicographer, should be taken to London to be touched by Queen Anne to be cured of "scrofula," as tuberculosis was then known. Dr. Floyer himself suffered from asthma, which prompted him to study the disease in detail and produce *A Treatise on Asthma*, which accurately described many aspects of the disease and gave the first detailed description of emphysema.

As his years began to pile up, Floyer became interested in the medical aspects of aging and produced *Medicina Gerocomica*, a text aimed at preserving old men's health. The preface begins with "every man is a fool or becomes a physician when he arrives at forty or fifty years of age." What he meant was that as the years creep along, people look for ways to retard the process of aging. He promoted physical exercise, but not too strenuous. Sailing, pruning trees, riding, bowling, billiards, ninepins, fishing, and walking were favored. Floyer also advocated temperance when it came to tobacco and alcohol, and interestingly he suggested that with age "instead of cold baths, hot water does sometimes have advantages." Although his emphasis on baths may not have much scientific backing, he did live to the age of eighty-five, very impressive for the time.

There is perhaps another explanation for his longevity, as was offered in a letter written by a friend with whom he visited just before his demise. "Sir John has all his memory, understanding, and all his senses good, and seems to labor under no infirmity." He continues with a possible explanation. "He had a wife, who was not the most easy or the most discreet; but he is of a happy temper, not to be moved by what he cannot remedy, which has I really believe in great measure helped to preserve his health and prolong his days." In other words, he didn't stress himself about things he could not control.

Dr. Floyer's legacy is a mixture of a gullible reliance on ancient authorities, remedies based on personal belief, and experimental investigation. His accurate measurement of the pulse under different conditions was pioneering work, but like most new ideas was criticized by the "establishment." Indeed, an 1820 text, *Advice and Maxims for Young Students and Practitioners of Medicine*, stated that "it is a pompous practice too much in fashion on all occasions to time the pulse by a fine gold watch, which, perhaps, may be thought to give consequence, like a gold-headed cane." (Gold-headed canes were awarded by the Royal College of Physicians to outstanding practitioners, and undoubtedly just the knowledge of being treated by such a doctor enhanced outcomes.)

But you can't keep a good idea down. In 1864, Irish physician Robert Graves, after whom Graves' disease is named, reintroduced the taking of the pulse using a watch as an important feature of clinical medicine and is often erroneously given credit for the development of the second hand on watches. That credit as we have seen belongs to Samuel Watson and Dr. John Floyer, who advocated accurate measurement of the pulse so that "we may know the natural pulse and the excesses and defects from this in diseases." To this day, the taking of the pulse is an important feature of clinical medicine.

ETHER FROLICS

The marble and granite statue in the Boston Common depicts a physician in medieval clothing holding a cloth next to the face of a man who seems to have passed out. An inscription on the base of the statue reads, "To commemorate that the inhaling of ether causes insensibility to pain, first proved to the world at the Mass. General Hospital in Boston, October A.D. 1846." No names are mentioned.

It was on October 16, 1846, that dentist William Morton ushered in the era of surgical anesthesia by putting printer Gilbert Abbot to sleep with fumes of ether from an inhaler he had devised. Surgeon John Collins Warren then proceeded to remove a tumor from the patient's neck without any of the usual screaming or thrashing about. Warren looked up at the doctors who had witnessed the event in the surgical theater that would become known as the "ether dome" and proclaimed, "Gentlemen, this is no humbug." The quote was in reference to a previous failed attempt by the dentist Horace Wells to demonstrate anesthesia with nitrous oxide, or laughing gas, at the same hospital. In that case, Wells hadn't waited long enough for the nitrous oxide to take effect, and the patient howled in pain as Wells attempted to extract a tooth. He exited in disgrace to the cries of "humbug."

Although Morton gets credit for the first organized demonstration of ether anesthesia, he certainly was not the first to experiment with the chemical. The sleep-inducing effect of ether was first recorded some three hundred years earlier when famed Swiss alchemist, philosopher, and physician Paracelsus noted that its vapors would induce a state of unresponsiveness in chickens. Ether does not occur in nature, so where did Paracelsus get it?

In 1540, German physician and botanist Valerius Cordus discovered that heating alcohol with sulfuric acid, then known as oil of vitriol, yielded a new highly flammable substance with a characteristic smell. Vitriol was the archaic name for compounds that today are termed sulfates. Cordus discovered that heating a solution of green vitriol, or iron (II) sulfate, a naturally occurring mineral, yielded "oil of vitriol." Then in the seventeenth century, German-Dutch chemist Johann Glauber found that burning sulfur with saltpeter (potassium nitrate) produced sulfuric acid. Potassium nitrate decomposes to yield the oxygen needed to convert sulfur to sulfur trioxide, which dissolves in water to produce sulfuric acid. In the nineteenth century, potassium nitrate was replaced by vanadium pentoxide that acted as a catalyst, allowing for easier production of sulfur trioxide. This was the method used to produce the sulfuric acid needed for the synthesis of ether in the 1800s.

Before ether's triumphant performance in 1846 at Massachusetts General, it had developed a reputation as a recreational substance. Middle-class partygoers and medical students in both Europe and America frolicked under the influence of ether. More curiously, drinking ether was common in Europe and was particularly popular in Ireland, where the Catholic Church promoted abstinence from alcohol and asked people to pledge not to drink alcohol. Drinking ether was a way to get around the pledge. Ether was sold in pubs and shops until the 1890s, when it was classified as a poison.

Dr. Crawford Long had taken part in ether frolics as a medical student at the University of Pennsylvania, and when he took over a rural medical practice in Georgia in 1841 he recalled that ether frolickers sometimes developed bumps and bruises of which they seemed to be oblivious. Could ether be used to relieve pain, he now wondered? The answer came when he delivered his wife's second baby with the aid of ether anesthesia. Long went on to perform a painless dental extraction, and in 1842 used an ether-soaked towel to put James Venable to sleep before proceeding to excise two tumors from his neck. But Dr. Long was not an academic, was not interested in publishing, nor did he crave fame or fortune.

It was two years after William Morton's celebrated demonstration that Long documented his efforts in the *Southern Medical and Surgical Journal* in a paper entitled "An Account of the First Use of Sulfuric Ether by Inhalation as an Anaesthetic in Surgical Operations." He described a number of cases, including the amputation of two fingers of a boy who was etherized during one procedure and not the other. Long reported that the patient suffered terribly without ether but was insensible with it. The reason he had waited to publish, he said, was the need to overcome criticism by local colleagues who had suggested that the ether effect was just an example of mesmerism, which at the time was promoted as a pain reduction method.

With his publication, Long added his name to the list of people claiming to have been the inventors of ether anesthesia. There was William Morton of course, and Charles Jackson, a physician who had given up medicine to establish a private laboratory for analytical chemistry where he also taught students, including Morton, who had come to expand his scientific knowledge. Jackson claimed that he had introduced Morton to ether anesthesia and the two were involved in a rancorous battle for years. There was also a Berkshire Medical College student, William E. Clarke, who claimed he had first used ether to put patients to sleep.

It was because of the controversy that the Boston monument does not bear the name of any of the claimants. But it does bear a biblical quote from Isaiah: "This also cometh forth from the Lord of Hosts which is wonderful and excellent in working," addressing the worry people had that relief of pain was somehow interfering with God's will. The quote suggests that medical intervention is itself a gift from God and is backed up by a relief on the statue depicting a woman who represents Science Triumphant sitting atop a throne of test tubes, burners, and distillers, with a Madonna and Child looking on with approval. There is also a Civil War scene on the side of the monument with a Union field surgeon standing ready to amputate a wounded soldier's leg. The soldier sleeps peacefully. Thanks to ether, he would feel no pain.

DEALING WITH THE PLAGUE

Thanks to the Ebola crisis the word "quarantine" is appearing with increased frequency in news reports and daily conversations. The term derives from *quaranta giorni*, meaning forty days, and traces back to the fourteenth century when the city of Dubrovnik, now in Croatia, was under Venetian rule. "The Great Pestilence" or "The Great Plague" as it was known at the time was devastating Europe and as a form of protection Dubrovnik declared that all ships and people had to be isolated for forty days before entering the city. Later the disease would be referred to as the "Black Death," probably because of the gloom it brought, although some theorize that the "black" referred to the terrible dark bruising of the skin due to internal bleeding, a hallmark of the disease. Between 1345 and 1360, the plague wiped out roughly half of Europe's population. The cause was unknown, but it was clear that the disease was contagious. Once it took hold, it spread like wildfire. In Milan, doctors advised that victims should be walled up in their homes along with healthy family members, a measure that apparently worked since Milan had the lowest death rate from the plague in all of Italy.

It would not be until 1894 that Alexandre Yersin of France's Pasteur Institute would identify a bacterium as the causative agent while investigating an outbreak of the plague in Hong Kong. The bacterium, eventually named *Yersinia pestis* in his honor, is thought to have originated in Asia, where it found a hospitable environment in fleas, insects that would readily transmit it through their bites. Since fleas infested rats and mice, rodents that were regular passengers on ships, the disease spread throughout the Mediterranean and Europe.

Infection with the bacterium can take several forms, with bubonic plague being the most notorious. This term originates from the Greek for "groin" due to the characteristic swellings of the lymph glands particularly in the groin, an area close to the legs where flea bites are most likely to occur. In septicemic and pneumonic plague, bacteria enter the bloodstream and can be transmitted from person to person, especially though the coughing associated with pneumatic plague.

When science fails to find an explanation for a phenomenon, superstition and quackery rush in to fill the void. And there certainly was no scientific explanation for the plague in the fourteenth century. The Church decreed that the Black Death was punishment for human sin. Lepers, because of their outward signs that resembled the plague, were blamed, as were astrological alignments and volcanic eruptions. "Flagellants" believed God's punishment could be avoided by stripping to the waist and whipping themselves as they marched from town to town. Jews were also targeted, accused of poisoning wells. Many Jewish communities in Europe were exterminated in hopes of bringing an end to the plague. In Cologne, thousands of Jews were burned alive after being accused of starting the plague. Black cats also became victims. They were thought to be witches in an animal form, casting their spell on the population. Since cats were a natural enemy of the disease carrying rats, hunting them actually increased the spread of the plague.

As far as treatments went, there were none. Since the plague was often accompanied by a terrible smell, people walked around with flowers under their noses, hoping to ward off the stench and the disease. This of course did nothing. Neither did the burning of aromatic woods to purify the atmosphere. Other attempts to remedy the "sick air" included the ringing of bells and the firing of guns. Birds were released indoors so that the flapping of their wings would break up the pestilence. Bathing was thought to be dangerous, as was the consumption of olive oil. And one of the most bizarre pieces of advice given to men was that if they valued their lives, they must preserve their chastity. Apparently no such advice was given to women.

The belief that pleasant smells were of some help persisted through the seventeenth century when the Great Plague once again terrified Londoners. The classic children's rhyme about a "pocketful of posies" is said to date back to that time, although that is somewhat questionable, since the first publication of the poem was apparently in the 1880s. Posies were flowers, but as the lyrics indicate, they did not do much good against the "ring of rosies," the rose-colored rash in the form of a ring around flea bites. The outcome of the disease was clear: "atishoo, atishoo, we all fall down." And some 100,000 citizens of London did. Holding garlic in the mouth, swishing vinegar, or burning sulfur to get rid of the "bad air" did no good. Smoking was also thought to be protective, and even children were forced to smoke tobacco, with threats of being whipped if they didn't.

Cases of plague still occur today, but they are rare. The first effective treatment appeared in 1932 with the advent of the sulfonamide drugs, but today the standard treatment is antibiotics such as streptomycin, chloramphenicol, tetracycline, and the fluoroquinolones. Unfortunately, the possibility of using the bacterium as a form of biological warfare exists. Indeed, recognition of the contagious nature of the plague resulted in the first example of biological warfare when in a 1347 attack on the Crimean city of Caffa, the Mongols catapulted the bodies of plague victims over the city walls. More recently, in 1940, a Japanese plane dropped a load of infected rat fleas over a Chinese town, causing a local plague. Today, stories circulate about various countries having developed bacterial warfare agents in the form of strains of the bacterium that are resistant to all drugs.

But for now our major worry is the Ebola virus and quarantine is the most effective way to halt its spread. In this case, about twenty-one days after exposure to an infected person is sufficient, that being the incubation period for the disease. If no symptoms appear after this period, there is no worry about the infection being passed on. It appears that contagion occurs only when symptoms are present. But if quarantine isn't instituted when appropriate, we may have to confront a scourge that will outdo the Black Death.

LAUDING MORPHINE

Famed Harvard Medical School professor and writer Oliver Wendell Holmes opined in the 1800s that if all medicinal drugs used at the time could be sunk to the bottom of the sea, "it would be all the better for mankind and all the worse for the fishes." He was, however, careful to make an exception for opium, the resinous latex that exudes from the seedpod of the opium poppy. The use of opium predates written history, with remains of poppy pods having been found in caves that are known to have been occupied by humans as early as 10,000 B.C. Of course, exactly what role the poppies played in the cavemen's lives is not known, but it is possible that by trial and error they chanced upon the calming and pain-relieving properties of poppy juice. By 3500 B.C., the Sumerians, who occupied Mesopotamia, now western Iraq, were trading opium with other civilizations, implying widespread awareness of the effects of consuming opium. Ancient Roman and Greek physicians prescribed opium for melancholy, pain, coughing, and diarrhea, conditions for which opium does provide relief.

"Take opium, mandragora, and henbane in equal parts and mix with water," a twelfth-century treatise advised doctors. "When you want to saw or cut a man," it continued, "dip a rag in this, put it to his nostrils, and he will sleep so deep that you may do what you wish." The problem was that sometimes the sleep became permanent. In the sixteenth century, Swiss-German physician and alchemist Paracelsus, perhaps best known for his dictum "only the dose makes the poison," discovered that opium was more soluble in alcohol than water and recommended a solution he called "laudanum" for relieving pain. The name came from the Latin *laudare*, meaning "to praise," but laudanum wasn't always praiseworthy. Determining appropriate dosage was difficult because solutions varied greatly in concentration of active ingredients. Physicians also began to notice that stopping the drug after long-term use led to "great and intolerable distresses, anxieties, and depression of the spirit," essentially the first reports of addiction and withdrawal.

Laudanum became the Victorian era's most popular medicine. It was of course used for pain, but many looked to it for its euphoria-inducing effect. This was well-described in Thomas De Quincey's classic work *Confessions of an English Opium Eater*. De Quincey had been introduced to laudanum in 1804 as a treatment for trigeminal neuralgia, a disease of the trigeminal nerve that runs down the face. Even mild stimulation such as brushing the teeth or exposure to the wind can trigger a jolt of excruciating pain. Little wonder that De Quincey sang the praises of laudanum: "Here was a panacea for all human woes, here was the secret of happiness." Samuel Taylor Coleridge was also keen on laudanum. His famous poem "Kublai Khan," about the thirteenth-century Chinese Emperor, was based on a dream he had while in a laudanum-induced stupor. For those of you interested in trivia, Mary Todd Lincoln was a laudanum addict, and so was Mattie Blaylock, common-law wife of Wyatt Earp. In *Uncle Tom's Cabin*, Cassy kills one of her children with laudanum to prevent him from growing up in slavery, and Bram Stoker's Dracula puts Lucy's maids to sleep with laudanum before sinking his teeth into their necks.

In the early nineteenth century, German pharmacist Friedrich Wilhelm Sertürner became the first person ever to isolate an active ingredient from a medicinal plant, naming the compound he managed to purify from an opium extract "morphine" after Morpheus, the Greek god of dreams. After experimenting on mice and stray dogs, he took it himself and enlisted friends to ingest the substance to determine an appropriate dosage for humans. Sertürner noted that 30 milligrams induced a happy, lightheaded sensation, a second dose caused drowsiness and fatigue, and a third dose caused a deep sleep with nausea and headaches upon awakening. After this, his friends refused to continue the experiment. The introduction of the hypodermic syringe made appropriate dosages easier to administer, but also facilitated abuse. Morphine, as the drug was called in English, was used extensively during the Civil War for pain of battle wounds, and thousands of survivors became addicted to the drug.

It wasn't until the 1970s that the mechanism of morphine's activity was understood in detail. The molecule binds to receptors in the central nervous system, interfering with the transmission of pain signals. But there was the question as to why such receptors exist. After all, the human body did not evolve to respond to extracts of a poppy that grows in Asia. As it turns out, morphine's molecular structure happens to mimic natural pain relievers produced by the body, compounds that were appropriately termed endorphins from "endogenous" and "morphine."

Over the years, attempts have been made to separate morphine's medicinal properties from its addictive ones by modifying its molecular structure. Diacetylmorphine was introduced by the Bayer company in 1894 as "Heroin" because in clinical trial subjects experienced a "heroic" feeling. It was believed to be more potent and less addictive than morphine and was actually used to treat morphine addiction. That didn't work. Actually heroin is more fat-soluble than morphine, allowing it to move easily across the blood-brain barrier into the brain where it converts to morphine. This means a greater supply of morphine to the brain and a more powerful effect than what is achieved by administering morphine, along with a greater potential for addiction. By the 1920s, heroin was banned, giving rise to an illicit industry to convert morphine to heroin, an industry that is ten times greater than the production of medicinal morphine. There are no viable synthetic methods to produce morphine, so the drug is still extracted from the poppy, although preliminary experiments have indicated a possibility of production using genetically engineered yeast. Morphine is the proverbial double-edged sword. It can be a pain-alleviating angel or an addiction-causing devil.

THE POWER OF HEAT

The place was Edinburgh, Scotland. The occasion, the Edinburgh Science Festival. There were a number of captivating presentations, but my biggest thrill came from looking out the hotel window. A light rail track was being constructed just outside, and the workers were busy welding. My eyes popped when I saw what they were doing. I was looking at a live thermite reaction! I had talked about this reaction in class on numerous occasions and marveled on it in videos but had always deemed it too dangerous to perform.

A chemical reaction that produces heat is said to be exothermic. The most common example would be the combustion of a fuel. Light a candle and you can feel the heat that is produced. The hottest part of a flame, where the color is a light blue, can reach a temperature of about 1,400°C. But that is a low temperature compared with 2,500°C produced by the thermite reaction between aluminum and iron oxide. Essentially this reaction involves the transfer of oxygen from the iron oxide to aluminum to yield aluminum oxide and metallic iron. At this high temperature the iron is in its molten form and sets fire to any combustible material in its path, making the thermite reaction ideal for use not only in welding but also in incendiary bombs and grenades.

Back in 1893, German chemist Hans Goldschmidt was looking for a way to produce pure metals from their ores. The classic method for extracting iron relies on heating iron oxide ore with carbon. The carbon is converted to carbon dioxide as it strips oxygen from the iron, leaving behind metallic iron. Some unreacted carbon, however, tends to contaminate the iron. Goldschmidt was looking for a way to produce iron without the use of carbon and hit upon the reaction of iron oxide with aluminum. He was impressed by the remarkable amount of heat produced and suggested that the reaction he had discovered could be used for welding. In 1899, the thermite reaction was put to a commercial use for the first time, welding tram tracks in the city of Essen.

It didn't take long for the military to realize the potential of this extreme exothermic reaction in warfare. In 1915, the Germans terrorized England with Zeppelins dropping incendiary bombs based on the thermite reaction. By World War II, the battle was on not only between Allied and German armed forces, but also between their scientists and engineers who sought to produce more effective incendiary devices. The Germans came up with the Elektron bomb, named after Elektron, an alloy composed of 86 percent magnesium, 13 percent aluminum, and 1 percent copper that was used for the casing of the bomb.

This alloy burns with a very hot flame but requires a high temperature for ignition. The thermite reaction was up to the task. When an Elektron bomb hit the ground, a small percussion charge of gunpowder ignited a priming mixture of finely powdered magnesium and barium peroxide. This reaction produced the heat needed to ignite the thermite mix of aluminum and iron oxide, which in turn ignited the highly combustible casing. The Allies developed similar types of bombs, resulting in the most destructive air raid in history, which was not Hiroshima or Nagasaki, but the firebomb raid on Tokyo in March of 1945. An Allied bombing of Dresden the same year with incendiary bombs virtually destroyed the whole city. During World War II, the Allies dropped some thirty million four-pound thermite bombs on Germany and another ten million on Japan.

Thermite hand grenades were also used during the war to disable artillery pieces without the need for an explosive charge, very useful when silence was necessary to an operation. This involved inserting a thermite grenade into the breech of a weapon and then quickly closing it. The great heat produced by the thermite reaction welded the breech shut and made loading the weapon impossible. Alternatively, a thermite grenade was discharged inside the barrel of an artillery piece, making it useless.

During the Vietnam War, thermite grenades found a different use. From the start of hostilities, putting a crimp into the enemy's food supply was part of the U.S. military strategy. Since rice was a staple for the Viet Cong, destroying rice paddies was a primary goal. At first, attempts were made to blow up rice stocks and destroy paddies with hand grenades and mortars but this proved to be maddeningly difficult. The next idea was to burn the rice paddies with thermite grenades. All this did was scatter the rice grains, which could then still be harvested. Another approach was needed.

Enter Agent Blue, an arsenic-based herbicide, unrelated chemically to the more infamous Agent Orange. Agent Blue affects plants by causing them to dry out, and as rice is highly dependent on water, spraying Agent Blue on rice paddies can destroy an entire field and leave it unsuitable for further planting. The U.S. used some twenty million gallons of Agent Blue during the Vietnam war, destroying thousands of acres of agricultural fields and defoliating wooded areas that the Viet Cong used to ambush American troops.

Recently the thermite reaction made the news in a different context. Conspiracy theorists purport that it was thermite explosives planted inside the World Trade Center that brought down the twin towers in a CIA-coordinated plot. They also maintain that the moon landing was faked and that the U.S. government is hiding the bodies of aliens. Some also claim that the rise of Donald Trump was engineered by a Democratic conspiracy and that on the verge of being elected he would announce, "Fooled you." Wouldn't that be something? It would trump the thermite reaction for heat generated.

A BLOODY GOOD YARN

I've never been very fond of eating liver, but I do enjoy using it as a teaching tool. Drop some hydrogen peroxide on a piece and watch it foam! How does that happen? A foam forms when bubbles of a gas are trapped in a liquid or solid. In this case oxygen is generated when hydrogen peroxide breaks down into oxygen and water on contact with catalase, an enzyme found in liver. Enzymes are special protein molecules that speed up chemical reactions. But why should liver contain an enzyme that helps degrade hydrogen peroxide? Because hydrogen peroxide actually forms as a product of metabolism and can do some nasty things. It can break apart to yield hydroxyl radicals that attack important biochemicals, such as proteins and DNA. To protect itself, the body makes catalase, an enzyme that decomposes hydrogen peroxide before it can form hydroxyl radicals. Actually, the formation of hydrogen peroxide in cells is an attempt by the body to protect itself from an even more dangerous substance, superoxide.

Oxygen is a double-edged sword. We can't live without it, but it also hastens our demise by playing a role in the aging process. Here's what happens. Electrons are the "glue" that hold atoms together in molecules and all sorts of electron transfers occur between molecules when they engage in the numerous chemical reactions that go on in our body all the time. Sometimes during these reactions an electron is transferred to oxygen, converting it into a highly reactive superoxide ion that attacks and rips other molecules apart. But we have evolved a defense system, an enzyme called superoxide dismutase that gets rid of superoxide by converting it into hydrogen peroxide, and although hydrogen peroxide is potentially dangerous, it is less dangerous than superoxide. Still, it does present a risk, and this is where catalase enters the picture. It breaks the peroxide down into oxygen and water. And that is why hydrogen peroxide foams when poured onto liver.

If you have ever used hydrogen peroxide to disinfect a cut, you may have also noted some bubbling since blood can decompose hydrogen peroxide into oxygen and water. The catalyst this time is not an enzyme, but the heme portion of hemoglobin, the oxygen-carrying compound in red blood cells. Swiss chemist Christian Friedrich Schonbein, best known for his discovery of "guncotton" upon using his wife's apron to wipe up an accidental spill of nitric and sulfuric acids, was the first to note bubbling when hydrogen peroxide was mixed with blood. He reasoned that if an unknown stain caused foaming on treatment with hydrogen peroxide, it probably contained hemoglobin, and was therefore likely to be blood. Introduced in 1863, this was the first presumptive test for blood. But since hydrogen peroxide tends to decompose slowly by itself, looking for extra bubbles was a challenging endeavor.

A significant improvement was introduced in the form of the Kastle–Meyer test, which produces a color change in the presence of hemoglobin. This relied on the chemistry of phenolphthalein, well-known today to students as an acid-base indicator. Phenolphthalein is colorless in acid but turns a deep pink in a basic solution. In this case, though, the important feature is that phenolphthalein can be reduced with zinc into colorless phenolphthalin, which along with a base is present in the test reagent.

In the usual process, a drop of alcohol is added to an unknown stain to dissolve any hemoglobin that may be present, followed by rubbing with a swab that has been treated with the Kastle–Meyer reagent. A drop of hydrogen peroxide is then applied to the swab. If hemoglobin is present, the hydrogen peroxide decomposes to yield oxygen that in turn oxidizes the phenolphthalin to phenolphthalein. Since the solution is basic, a pink color develops, indicating the presence of blood. The test is very sensitive, but is not specific for human blood. Animal blood will also yield a positive reaction, as will oxidizing agents such as some metal ions.

This reaction of hydrogen peroxide with hemoglobin is also the basis of the "luminol" test used by crime scene investigators to detect traces of blood that may not be visible at all. The technique is to spray the suspect area with a solution of luminol and hydrogen peroxide. If blood is present, the peroxide will yield oxygen that then reacts with luminol to produce a blue glow. This reaction was first noted in 1928 by the German chemist H.O. Albrecht and was put into forensic practice in 1937 by forensic scientist Walter Specht.

Even dried and decomposed blood gives a positive reaction, with the blue glow lasting for about thirty seconds per application. The glow can be documented with a photo, but a fairly dark room is required for detection. The reaction is so sensitive that it can reveal blood stains on fabrics even after they have been laundered. In one case, a pair of washed jeans with no visible stains gave a positive test with luminol on both knees.

Neither the Kastle–Meyer test nor the luminol test can identify whose blood is involved, but once a stain has been determined to be blood, traces of DNA can be extracted and an identification carried out. In the example of the jeans, DNA analysis was able to exclude the possibility that the blood came from the owner of the jeans.

Luminol analysis does have drawbacks. Its chemiluminescence can also be triggered by a number of substances such as copper-containing compounds and bleaching agents. Had the jeans been washed with a detergent containing a bleaching agent, the blood would not have been detected. Criminals aware of this have been known to try to wash away traces of their crime with bleach. The result is that residual bleach makes the entire crime scene produce the typical blue glow, which effectively camouflages any blood stain. And if you want to see a really impressive glow, spray a piece of liver with a luminol test solution. Just don't eat it after.

MEMORIES OF LINUS PAULING

There are some memories that become indelibly etched in one's mind. It was a June day in 1980 at the Chemical Institute of Canada's annual conference that I found myself waiting to hear from Dr. Linus Pauling, one of the most famous scientists in the world. Pauling had already won a Nobel Prize in Chemistry in 1954 and one for Peace in 1962, making him the only person to have been awarded two unshared Nobels.

As I was forging my own career in chemistry, I had become fascinated with Pauling's breadth of contributions. His seminal ideas about chemical bonding, based on the arrangement of electrons in orbitals around atoms, are featured in every introductory chemistry textbook, and no biology text is complete without a reference to Pauling's demonstration of sickle cell anemia being caused by an abnormal protein. Indeed, sickle cell anemia was the first disease to be understood at the molecular level. Pauling also introduced the concept of electronegativity, a measure of an atom's affinity for electrons, making possible the prediction of the strength of bonds formed between different atoms. He made huge contributions to protein chemistry and almost solved the structure of DNA before Crick and Watson.

I was also intrigued by the story of how Pauling had become interested in chemistry. It happened at the age of thirteen when a schoolmate, Lloyd Jeffress, who would go on to become a noted psychologist, invited him over to see some experiments he had carried out with his chemistry set. Pauling was smitten and soon began his own experiments, in one instance concocting mixtures that would explode when a streetcar passed over them. That recalled my own tinkering with ammonia and iodine to make nitrogen triiodide, a chemical that explodes with a snap on slightest contact, even a touch with a feather.

There was another reason I was keen to hear Pauling speak. Although he was much admired in the scientific community for his contributions to chemistry and for establishing the field of molecular biology, he had received a fair degree of criticism for his 1970 book, *Vitamin C and the Common Cold*. In this little volume, he outlined his belief that high doses of vitamin C cured colds. The problem was that the famous scientist, who had published over 800 peer-reviewed research papers, based his pet theory on his personal experience. Nevertheless, spurred by Pauling's claim, a number of trials had been carried out by 1980, all failing to show a curative effect.

Pauling's enthusiasm for vitamin C did not stop with the common cold. He claimed that large doses of vitamin C were beneficial in the treatment of cancer, a notion that was put to test in various trials, with most studies concluding that the vitamin worked no better than a placebo. Pauling countered that the trials failed because oral instead of intravenous doses were used. This notion still holds sway with some researchers and with some nonmainstream clinics that offer vitamin C therapy for cancer. Pauling also promoted vitamins for the treatment of mental disease, and coined the term "orthomolecular medicine" for the treatment of disease with substances normally present in the body, a concept criticized by conventional practitioners. So by 1980, Pauling was both revered and scorned. I was anxious to hear his address.

The great man began by describing vitamin C's role in the

formation of collagen, the structural protein in connective tissue. Collagen is composed of protein strands linked together into a three-dimensional network through their lysine residues by the action of lysyl hydroxylase, an enzyme that requires vitamin C to function. A lack of vitamin C disrupts the links between collagen's protein strands, resulting in scurvy. At this point, Pauling declared that such damage to collagen is also the cause of heart disease. He explained that collagen strands in an artery break down in the absence of enough vitamin C and the liberated lysine residues then bind to the lipoproteins that transport cholesterol around the bloodstream. This leads to the buildup of plaque in arteries that can eventually rupture and trigger the formation of a blood clot that can cause a heart attack.

Then came a stunning moment. Dr. Pauling showed a graph demonstrating the decline of deaths from heart disease starting in 1970. He then overlaid a curve showing the increase in vitamin C supplement sales, probably thanks to his book published that same year, and astonishingly offered this as evidence that vitamin C was instrumental in reducing the risk of heart disease. Actually, the decline started some ten years earlier, but there was a bigger issue. Although I was not as tuned in to the difference between associations and cause and effect relationships as I am now, I remember thinking, "Whoa!" Just because heart disease deaths had declined and vitamin C sales had increased did not mean that there was a causative connection!

Curiously, I had just been talking in one of my classes about the nutritional value of frozen foods and the concern raised by the increasing sales of TV dinners due to the large doses of salt added to make up for the taste lost to freezing. As I was listening to Pauling, it occurred to me that one could just as well have made the argument that TV dinners reduced the risk of heart disease. As I pondered this, Pauling had already gone on to suggest that heart disease can be treated with large doses of vitamin C and lysine. And so on that day back in 1980, my scientific hero lost some of his luster.

In 1992, I heard him speak again. After revealing he had prostate cancer, a brave soul timidly asked why his daily dose of 18 grams of vitamin C had not offered protection. Without blinking, the ninety-two-year-old Pauling retorted that if had he not taken the vitamin, he would have been afflicted decades earlier. Who knows?

THE INTOXICATING SCIENCE OF WINE

I'm quite adept at turning water into wine. Just pour a colorless solution of ferric sulfate into a glass that has a bit of potassium thiocyanate at the bottom and, presto, water changes into "wine." An interesting little demonstration of the formation of a blood-red complex between ferric and thiocyanate ions. But this is nowhere as interesting as the chemistry of creating real wine.

I'm no oenophile. Frankly, I don't derive much pleasure from sipping wine. But I do find the science of wine and winemaking quite intoxicating. And what a complex science it is! We've been trying to figure out the details of fermentation, the second-oldest chemical process harnessed by humans (fire being the first), for thousands of years, but it refuses to give up all its secrets.

Here's what we know. Grapes are little chemical factories that use carbon dioxide from the air and nutrients from soil to produce an array of sugars, acids, and numerous polyphenols. They also provide a hospitable environment for various yeasts and bacteria that occur naturally in the air and ground. To make wine, just crush the grapes, allow the yeast on the skins to convert the sugars to alcohol, and then let the liquid sit around for a while as the bacteria release enzymes that catalyze a torrent of reactions, transforming the grape's chemicals into the literally thousands of compounds that eventually determine the wine's aroma and taste. Store the wine in oak barrels, and the complexity of the flavor will be further increased by substances extracted from the wood.

Since the composition of the grapes depends on the seed variety, soil quality, amount of sunshine, rainfall, average temperature, length of aging, and even the altitude at which they are grown, it is evident that the variety of wines that can be produced is almost infinite. Subtle differences matter. For example, more 3-isobutyl-2-methoxypyrazine, a compound with an unpleasant bell pepper–like odor, forms when grapes are in the shade rather than in direct sunlight. Simply pruning leaves from vines to expose grape clusters to more direct light can address the problem.

Any attempt to understand the intricacies of wine production, with an eye to improving vintages, must start with getting a grip on just what compounds may be responsible for the aroma and flavor. This involves some sophisticated chemistry as well as refined palates. Basically, a sample of wine is passed through a chromatography column packed with some adsorbent substance. The different components of the wine stick to the adsorbent to different extents and emerge from the bottom of the column at different times. The fractions are then subjected to analysis by mass spectrometry and nuclear magnetic resonance (NMR) spectroscopy, instrumental techniques that can reveal the molecular structures of the isolated compounds.

A group at the Technical University of Munich led by food chemist Thomas Hofmann subjected an Italian wine to such analysis and then had trained experts taste the different fractions. They narrowed down the flavor to a mix of some thirty-five compounds and the aroma to another thirty volatiles. Eventually the researchers concluded that there are about sixty key aroma and taste molecules that when properly blended can simulate the taste and feel of any wine. What makes one taste like merlot and another like cabernet sauvignon is the difference in concentrations of these compounds.

A California enterprise, Ava Winery, is exploring the possibility of using the chemical information that has been gathered to make synthetic wine without grapes. The idea is that blending the right chemicals in the right concentrations can eliminate the expensive process of growing grapes and fermenting their juice. As one might expect, wine lovers in general are reviled by the idea of synthetic wine, the smell of which has been described as "that of the inflatable sharks one finds at a pool" and its aftertaste as "essence of plastic bag."

In China, where the wine industry is growing by leaps and bounds, researchers have come up with a different method to save on the cost of aging wine. Xin An Zeng, a chemist at South China University of Technology in Guangzhou, has shown that passing young wine through pipes surrounded by strong electric fields can alter its composition and, under the right conditions, can mimic the effects of aging. Unlike the unsupported claims that taste can be improved by placing rings of magnets around the neck of wine bottles or by resting wine glasses on magnetic coasters, experts have actually been able to detect differences in taste and aroma in wine subjected to electric fields. More significantly, chemical analysis has confirmed changes in composition.

The size of the field and length of exposure is important. In this case, bigger is definitely not better. Exposure for three minutes to a 600 volt per centimeter (V/cm) field was the best. Upping the field to 900 V/cm resulted in poorer taste. If anyone is thinking that this evidence for electric fields causing chemical reactions can be used to support the notion that cell phones can harm health, it should be pointed out that fields generated by cell phones are around 0.05 V/cm.

In any case, wine aficionados will never take kindly to such artificial treatments, preferring to discuss how natural processes make for wines that are "flamboyant," "barnyard," "flabby," "opulent," or "cigar-box-like." I wish I could relate to such expressions, but it seems my palate can't distinguish the Two Buck Chuck from the Il Barone I had the chance to try at the famous Castello di Amorosa winery in Napa. This costly wine has been described by an expert as "stunning, rich, and full-bodied with sweet tannin, a hint of smoked meats, deep fruit, and an impeccable balance." Meaningless to me, but I do admit to being intrigued by the possibility of a chemical connection between Il Barone and smoked meat.

CRYSTALLOGRAPHY SHEDS LIGHT ON MOLECULAR STRUCTURE

What pops into your mind if I mention the word "crystal"? My crystal ball tells me that it is likely to be a wine glass or perhaps a chandelier. "The secret of life" is probably not a candidate. But let me make a case for it.

That wine glass or chandelier? Not made of crystal at all. In fact glass is the very opposite of a crystalline substance. All matter is composed of atoms, molecules, or ions, and if these are arranged in an ordered pattern that extends in three dimensions, we have a crystal. Salt, rock candy, and diamond all feature a regular repeating pattern of constituents and are crystalline, but glass is amorphous, meaning that its component silicon and oxygen atoms do not have any long-range order. Crystals have sharp melting points and break along definite cleavage planes, while amorphous substances have a wide melting range and shatter randomly. Why then is crystal glass so named? Because it sparkles like diamond, a classic crystal. This is achieved by incorporating lead oxide or salts of barium or zinc into the glass to change the index of refraction, a measure of the degree to which a substance can bend a beam of light.

The notion of bending light is germane to our story about crystals and the "secret of life." It begins with Wilhelm Röntgen's famous 1895 discovery of a mysterious form of radiation he named "X-rays," using the standard mathematical term "X" for "unknown." Although others had earlier noted this mysterious form of radiation emanating from a vacuum tube in which a high voltage current caused electrons to travel from the cathode to the anode, it was Röntgen who first documented the effect he was unable to explain.

Were X-rays some form of invisible light, wondered the German physicist Max von Laue. The bending of light waves as they passed through a crystal was already well-known. Would X-rays behave the same way? Laue proceeded to project a narrow beam of X-rays onto a crystal of copper sulfate surrounded by photographic plates susceptible to exposure by X-rays. He noted that the rays were deflected, proving that they traveled through space in the form of waves, very much like light. The diffracted waves formed a pattern on the photographic plate, the specifics of which Laue was unable to interpret. But where he faltered, the father and son team of William and Lawrence Bragg succeeded.

The pattern, they proposed, was the result of the reflection of X-rays by planes of atoms in the crystal. Conversely, the Braggs suggested, from such an X-ray diffraction pattern, the location of atoms or ions in a crystal could be determined. If the atoms were joined to form molecules, the exact structure of these could also be determined.

To propose an analogy, consider the shadow cast by shining a beam of light on an object of unknown shape placed in a dark room. Moving the light source will change the shape of the shadow. Assembling all the shadow images would then allow for a determination of the three-dimensional shape of the unknown object. And that is the simplified basis of X-ray crystallography, one of the most powerful analytical techniques ever developed. Getting a handle on the exact molecular makeup of substances is the key to predicting chemical and biological behavior.

That brings us to the "secret of life" connection, which in turn brings us to DNA and its famous double helix molecular structure. Determination of that structure was fundamental to unraveling the mysteries of genetics, to the introduction of the era of recombinant DNA technology, and to the potential replacement of chemotherapy by targeted genetic manipulation. The names of Francis Crick and James Watson are intimately associated with the discovery of the structure of DNA since it was they who first used balls and sticks to build an accurate three-dimensional model of the molecule. Less well-known is that Crick and Watson shared the Nobel Prize for their discovery with physicist Maurice Wilkins, who had the original idea to study DNA by X-ray crystallographic techniques and whose lecture on the subject got Watson hooked on DNA research.

Even less well-known is the critical role played by Rosalind Franklin, Wilkins's colleague at King's College in London. It was Franklin who produced the famous "Photograph 51," an X-ray diffraction photo of the DNA molecule that turned out to be the key to building Watson and Crick's molecular model. Wilkins had shown the photo to Watson without Franklin's permission, which caused a fair degree of friction between the two. Franklin died of ovarian cancer in 1958 at the young age of thirty-seven, four years before the awarding of the Nobel Prize to Crick, Watson, and Wilkins. She could not be nominated for the Prize, which according to Nobel's stipulation cannot be awarded posthumously.

The Braggs had earlier been recognized with the 1915 Nobel Prize in physics "for their services in the analysis of crystal structure by means of X-rays," and to this date are the only father and son team to ever share a Nobel Prize. Max von Laue preceded the Braggs, receiving the 1914 Nobel "for his discovery of the diffraction of X-rays by crystals" and it is the centenary of Laue's award that prompted the United Nations to declare 2014 as the International Year of Crystallography. It is also happens to be the fiftieth anniversary of the Nobel Prize awarded to Dorothy Hodgkin for X-ray determination of the structures of important biomolecules such as penicillin and vitamin B12.

My crystal ball tells me that there will be more Nobel Prizes to come for crystallography given its potential for determining the structure of cell membrane proteins that are involved in numerous biological functions and that serve as targets for future drugs.

And if you are wondering, I really do have a crystal ball, although it happens to be glass. Real crystal balls made of quartz, a mineral that like glass is composed of silicon and oxygen but with the atoms being locked in an ordered arrangement, are available. They are made by taking a large piece of quartz, cutting it into an approximately spherical shape, and grinding it in a cylindrical container with abrasives of increasing fineness. I've got to get me one of those. It still won't let me peek into the future, but it will honor crystallography.

BRUSHING UP ON TOOTHBRUSH HISTORY

Where can I find a "natural" toothbrush? That was the question posed by a worried lady who did not want her mouth exposed to the "toxic chemicals" she believed may be leaching out of a plastic toothbrush. After I explained that nothing significant leaches out of the nylon bristles or the polyethylene or polypropylene handles, I facetiously remarked that toothbrushes do not grow on trees, so one would be hard pressed to find a "natural" version. It turns out that this is not exactly correct. There is in fact a "toothbrush tree," botanically known as *Salvadora persica*, the roots of which have a long history of use as a "natural" toothbrush in Asia and Africa, possibly dating back to 5000 B.C.

The roots of *Salvadora persica* can be cut into short twigs that are commonly referred to as "miswak." Soaking these in water and then scraping the bark off one end allows the wood fibers to be separated, essentially creating a brush. Once the fibers deteriorate, they can be snipped off and another segment of the bark can be scraped to expose a fresh brush. Miswak sticks may even have an effect that goes beyond cleaning the teeth through simple abrasion. The root exhibits antimicrobial activity that may help control the bacteria responsible for dental decay, with some studies actually showing a reduction in dental plaque with the use of miswak. Brushing with this natural root is especially popular in the Muslim world since the Prophet Muhammad is said to have recommended the practice in order to purify the mouth before prayers. Miswak sticks can be purchased today, even on Amazon, so I erred in stating that toothbrushes do not grow on trees.

Today's toothbrushes, however, are the products of a sophisticated plastics industry, but their shape resembles the first toothbrush ever produced, way back around 700 A.D. during the Tang dynasty in China. Bristles plucked from the neck of a hog were inserted into a slot in a bamboo shoot and secured with twine. Europeans, if they paid any attention to the cleanliness of their teeth at all, rubbed them with rags sprinkled with salt or soot. Then in 1780 in England, rag merchant William Addis "reinvented" the toothbrush and made it a commercial success.

According to accounts, which may or may not be true, the idea came to him while in jail, convicted of having started some sort of street riot. While languishing in his cell with a bad taste in his mouth, he noticed a broom in the corner and got the idea that a miniature version might be just the right instrument for removing food particles stuck between the teeth. After his release, he began to experiment with horsehairs or pig bristles inserted into holes drilled in a piece of bone. The European version of the toothbrush was born! That came just in time, as sugar from the West Indies was beginning to flood England, exacerbating the problem of dental decay. By the 1860s, the Addis company had introduced an automated manufacturing system, making mass production of toothbrushes possible. It was around that time that American ingenuity entered the picture, with H.N. Wadsworth filing a patent for a toothbrush with bristles of different lengths in angled clusters designed to contact all tooth surfaces. But a competitor soon appeared on the scene.

Dr. Scott's "Electric Toothbrush" promised not only to clean teeth but to "cause a current to flow into the nerve cells and roots of the teeth, and, like water poured upon a plant, invigorate and vitalize every part, arresting decay, building up and restoring the natural whiteness of the enamel, imparting pearly teeth and healthful rosy gums to all using it." If that sounds like quackery, it was. "Dr." Scott was not a doctor of any kind, and the brush was not electric. The handle did have a magnet embedded in it, which Scott claimed was "permanently charged with an electro-magnetic current." Pure nonsense of course, but apparently convincing to people whose knowledge of electrical generators was limited to magnets somehow being involved. Besides being a peddler of quack electrical toothbrushes, George Augustus Scott also sold an "Electric Flesh Brush" that was guaranteed to treat "Nervous Debility, Gout, Lumbago, Neuralgia, Toothache, Lameness, Impure Blood, and Impaired Circulation." It also would "beautify the complexion and impart vigor and energy to the whole body." It did none of these, but it was very effective at swelling "Dr." Scott's bank account.

Broxodent, the first truly electric toothbrush with a vibrating head, was introduced in 1959, followed by General Electric's more convenient product that featured a rechargeable battery. By this time, much to the relief of the hog population, pig bristles had been replaced by nylon, discovered by DuPont chemist Wallace Carothers in 1938. Nylon fibers also held their shape better and dried more quickly than natural bristles, reducing the chance of bacterial contamination.

Most people have appreciated these advances, as evidenced by surveys in which participants selected the toothbrush as the number one invention Americans could not live without, ahead of the car, the computer, the microwave oven, and the cell phone. Some, however, are not thrilled with the technological advances and hark back to a more carefree "natural" era, which of course only ever existed in their imagination. Manufacturers, though, will cater to people's whims, so that toothbrushes with a bamboo handle, albeit with nylon bristles, are available, as well as brushes with boar bristles, albeit with a plastic handle. But the lady who was worried about plastics can now purchase a "plastic-free" toothbrush from a company amusingly called "Life

Without Plastic." It has a handle made of sustainably harvested beechwood and pig hair bristles. It isn't vegan. Or kosher. And the pig bristles are imported from China, which may raise some questions. I think I will stick with my nonnatural, all plastic, scientifically shaped, soft nylon bristle toothbrush.

SORTING OUT STARCHES

The term "resistant starch" crops up with increasing frequency in nutritional discussions. It isn't that starch resists being eaten. But it does resist digestion, which is almost as good. Starch, the main component of foods such as potatoes, rice, bread, and pasta, is made up of glucose molecules joined in a chain. There are two basic varieties: amylose, in which the glucose units are linked in a straight chain, and amylopectin, in which shorter branches of glucose stem from the main chain. All these starch molecules congregate into granules that vary in shape and size.

Both amylose and amylopectin are broken down to glucose by enzymes during digestion, but not with equal ease. Amylose is digested with greater difficulty because the straight chains pack together tightly and are less exposed to digestive enzymes. In other words, it is more resistant to digestion. But the size and shape of the granules formed by the congregation of the starch molecules also determine the extent to which digestive enzymes can penetrate and break down the starch.

Resistant starch, instead of being broken down to glucose in the small intestine, travels through to the colon as if it were fiber, the indigestible component of plants. In the colon, some species of bacteria find the resistant starch to be a tasty morsel and use it as a source of food. Like us, bacteria also defecate, and the end result of their consumption of resistant starch is the release of short-chain fatty acids, butyric acid for example, that are believed to play a role in keeping the cells that line the colon healthy, reducing the risk of cancer. But a more significant feature is that since the starch resists digestion as it passes through the small intestine, less glucose is absorbed into the bloodstream. That means a reduced calorie intake and reduced need for the pancreas to release insulin.

One way to increase the resistant starch content of pasta is through appropriate cooking techniques. Heating, cooling, and then reheating starch increases the amount of amylose relative to amylopectin and also alters the nature of the starch granules, making them more resistant to digestive enzymes. In a study in the U.K., ten fasted subjects were randomized to eating either hot, cold, or reheated pasta on different days. On day one, the participants had freshly cooked pasta with a tomato sauce, on day two, it was cold pasta that had been chilled overnight, and on day three, they were served pasta that had been chilled and then reheated. Blood samples were collected every fifteen minutes for two hours.

Cold pasta led to a smaller spike in blood glucose and insulin than eating freshly cooked pasta. Surprisingly, cooking, cooling, and then reheating the pasta reduced the rise in blood sugar levels by 50 percent with a corresponding reduction in insulin release. This can be important for diabetics as well as for people wanting to watch their weight. And significantly, there is no downside to cooking, cooling, and reheating pasta. It tastes just as good.

While pasta may be a favorite in Europe and America, rice is the main source of starch in Asia and the Caribbean. Short-grain white rice in particular presents a problem because it is high in amylopectin. The consequence is rapid release of glucose into the bloodstream and a high glycemic index, which is linked to a higher risk of diabetes. This type of rice is the most popular in Asia because the amylopectin leaches out of the grains during cooking, absorbs some water, and forms a gel that holds the grains together so they can be more easily picked up with chopsticks. Long-grain rice by contrast has the most amylose and cooks up fluffy and nonsticky and has a lower glycemic index.

Just as cooking techniques can alter the glycemic index of pasta, they can also change the chemical profile of other starchy foods. Mashing potatoes, for example, increases their amylopectin content and frying rice reduces the amount of amylopectin. But most rice is consumed steamed. Nevertheless it is possible to reduce its amylopectin content.

Researchers in Sri Lanka have found that a relatively simple technique can impair the digestibility of rice, reducing its calorie content by some 10 to 15 percent. It was achieved by adding coconut oil, about 3 percent of the weight of rice being cooked, to boiling water before adding the rice. The cooked rice was then cooled for twelve hours in the refrigerator. What the oil actually does isn't totally clear, but the guess is that it forms a complex with amylose that prevents it from being attacked by the enzymes that normally would degrade it to glucose. All this is very interesting in terms of chemistry, but altering the ratio of amylose to amylopectin in rice in the North American diet, where rice is not a daily staple, is not likely to have a significant impact on health.

What about consuming brown rice, which has lots of fiber that prevents the absorption of starch? There is an issue here as well. Some varieties of brown rice have relatively high doses of arsenic absorbed from the soil by the grain's outer layer. It is this outer layer, the bran, that is removed in producing white rice. Brown rice has 80 percent more inorganic arsenic, the worrisome variety, than white rice, with the arsenic content depending on where the rice was grown. Brown basmati from California, India, or Pakistan has the lowest arsenic content. "Organic" rice has the same issue with arsenic as the conventionally grown variety since arsenic uptake has nothing to do with the use of pesticides or fertilizers. Just to be on the safe side, infants should not be given rice cereals more than once or twice a week. For adults, the arsenic issue in rice in the context of North American consumption patterns is not significant.

Of course, another alternative to reduce worries about pasta or rice is to replace them with more veggies.

GETTING DOWN TO EARTH

It seems the Earth is a giant pill. It can cure disease. Luckily you don't have to try to swallow it. But you do have to swallow some pretty bizarre "science" about its supposed curative effect. It's all about the Earth being a source of electrons, and disease being some sort of "electron deficiency." The thesis is that the reason we see so much chronic disease these days is that we wear synthetic-soled shoes and walk on carpets that insulate us from the ground. Apparently, leather soles are not so bad because sweat from the feet permeates the leather with moisture and body salts so the shoe becomes a semiconductor, permitting the body to receive electrons. Living or working in high-rises exacerbates the problem because we are further removed from the Earth's supply of electrons. And things are made even worse by the invisible electromagnetic fields generated by cell phones, computers, appliances, and wiring in the walls, all of which contribute to the body's positive electrical charge and its need to neutralize this with a flow of electrons.

The cure is "Earthing," a process that allows the free flow of electrons between the ground and our body. How do we facilitate this? Simple enough. Just walk barefoot, preferably on damp ground or moist grass since water is a great conductor. Seawater is especially good because its mineral content enhances conduction. So swimming in seawater, dangling your feet in it, or walking on a sandy beach are great ways to ground yourself. Who says? Dr. Joe Mercola.

If you haven't heard of Mercola, you have not been surfing the waves of questionable health advice on the web. He is an osteopathic physician whose practice now is limited to offering mostly iffy medical guidance on his popular website and selling a variety of dubious products. Apparently, though, he still occasionally uses a stethoscope because he claims he can hear "electrical chatter" that he attributes to the nervous system moving electrons about. I bet physicians get a charge out of that claim.

Needless to say, Mercola has an answer for those of us who run around in shoes all day, don't work underground, and use cell phones and computers. He suggests sleeping on special mats that allow the free flow of electrons through a grounding wire and exercising on yoga mats of a similar design. These wonder products are of course available from Dr. Mercola's website.

Mercola does not claim to have discovered this simple solution to complex health problems, one that apparently has been missed by the mainstream scientific community. His authority is Clinton Ober, who according to Mercola made the discovery that "could end up as groundbreaking as the germ theory." What sort of background does Mr. Ober have to shake the world in this fashion? It seems that at the age of fifteen he had to forgo school to take care of the family farm. Eventually he forged a career in the cable television business, but in 1993 decided to change his life after almost dying from some illness. For four years, he traveled around the U.S. in a forty-foot bus until an epic moment occurred as he stared across a bay in Florida.

"As I was asking myself what I should be doing, I automatically wrote on a piece of paper 'become an opposite charge, status quo is the enemy.'" The Earth, he says, was trying to send him a message that "in the modern electrical world with our bodies insulated from ground contact, we are vulnerable to electrical interference as our cells all transmit and receive the vital information that keeps us alive and healthy." This vulnerability could be countered, Ober determined, "by simple contact with the ground to neutralize charges in the body and thus protect the nervous system and the endogenous fields of the body from extraneous electrical interference."

Before long, Ober found the evidence he needed. He was looking at some studies that claimed to have found adverse effects in humans on exposure to electromagnetic radiation but were deemed to be inconclusive because the effects could not be reproduced in animals. He had hit pay dirt! The reason that sheep and baboons had not experienced the same adverse effects as humans, he reasoned, was because the animals were not wearing shoes and were not sleeping in comfortable beds insulated from the earth. They were naturally grounded! Mercola buys this argument and even contributes his own wacky insight. "Animals that live in the wild are not bothered with inflammation, cardiovascular disease, diabetes, arthritis, or even plaque on their teeth. This is why your dog or cat will crawl under the porch and lie on the bare earth if he isn't feeling well." I figure snakes must therefore be an especially healthy species. Maybe that is why snake oil is so popular.

Selling snake oil always works better if some reasonably sounding scientific mechanism is provided. In this case, it is all about the free radicals that form in the body as metabolic by-products and are implicated in a variety of disease processes as well as in aging. Free radicals are electron deficient and do their dirty work by stealing electrons from other molecules, such as DNA, damaging them and triggering disease. According to "Earthers," contact with the ground allows electrons to enter through the foot and satisfy free radicals' hunger for electrons, thus preventing damage to important biomolecules. Silly stuff. As silly as the original argument that walking with synthetic-soled shoes on carpet insulates us from the ground. Actually, walking on a wool carpet with rubber-soled shoes results in the shoes stealing electrons from the wool and transferring them to the body. Just touch a metal doorknob and watch the electrons jump from the body. That should spark some curiosity about the claims of Earthing being grounded in science.

BARKING UP THE RIGHT TREE

Otzi carved his longbow from the wood of the yew tree, intending to kill prey, and perhaps his enemies. He would never have dreamed that the same tree would one day yield a chemical that saves lives. Otzi, as he was named by researchers who examined his mummified body, found in 1991 in the Otzal Alps of Austria, died about 5,300 years ago. His well-preserved body, clothing, and tools offer a glimpse into what life was like in the Chalcolithic age. That term derives from the Greek for copper, and lithos for stone, describing a period before the Bronze Age when the addition of tin to copper was found to form an alloy harder than pure copper. Beside his longbow, Otzi also had an axe with a copper head and a handle made of yew wood.

The wood of the English or European yew, as the tree is

now called, has just the right springiness for the production of longbows. Otzi apparently knew about the value of yew, but it was the success of the English during the Hundred Years' War with the French that made yew wood bows famous. By that time, most of the yew wood was actually imported into England from Europe because a shortage had developed due to the use of longbows in the numerous battles the English fought among themselves. To make sure that enough longbows could be produced to satisfy the English penchant for war, The Statute of Westminster was passed in 1472, declaring that every ship coming to an English port had to bring four bow staves for every ton of cargo.

The dominance of the bow in battles began to wane with the introduction of gunpowder in the fifteenth century. The last battle in England to be fought mostly with the longbow was the Battle of Flodden in 1513 between the English and an invading Scottish army under King James IV. Artillery was used by both sides, but it was the bow and arrow that decided the battle, decisively won by the English with James himself being killed by an arrow. A curious aspect of that battle was adherence to the medieval code of chivalry with King James sending a notice to the English a month in advance of his intent to invade. This gave the English army, commanded by the Earl of Surrey, plenty of time to prepare.

Yew trees are often found around churches in England, with one popular explanation being that the roots are so fine that they can grow through the eyes of the dead and prevent them from seeing their way back to the world of the living. Another story has the trees being planted around churches to prevent animals from grazing on holy ground by poisoning them. It is true that all parts of the English yew, except the fleshy part of its berries, are poisonous, and there are cases where cattle and

patent three weeks later, although Goodyear claimed that he had actually discovered vulcanization in 1839. Since childhood he had been fascinated by rubber and was aware of the limitations of this strange exudate of the rubber tree. While it had a number of obvious uses based on its impermeability to water, it was saddled with the classic problems of becoming hard and brittle in the winter and soft and gooey in the summer.

Goodyear was determined to solve this problem and tried mixing rubber with various substances, having little success until he met Nathaniel Hayward, who, while working for a rubber company, had discovered that spreading sulfur on rubber eliminated its stickiness. Goodyear bought Hayward's patent and began to add sulfur to his own concoctions. Then one day he accidentally dropped some rubber treated with sulfur on a hot stove, and on trying to clean up the mess noted that the material was stretchy and refused to harden with cold or soften with heat. Eventually the term "vulcanized" was used to describe this newfangled rubber, after Vulcan, the Roman god of fire.

Goodyear of course did not realize it, but the heat had allowed sulfur atoms to crosslink the long chains of rubber molecules, yielding novel properties. Then in 1905, George Oenslager found that the addition of thiocarbanilide accelerated the vulcanization reaction, making it more economic. This spurred the development of a whole range of "accelerators" that control not only the speed of the reaction but also the number of sulfur atoms in the crosslinks. This in turn controls the properties of the finished product, determining whether it will be used for tires, shoe soles, saxophone mouthpieces, hockey pucks, or condoms.

As is often the case, an advance brings a "but." In this case it is the possible formation of nitrosamines from the accelerators during the vulcanization. This has caused concern because nitrosamines are carcinogenic, but a study that considered the use of fifty condoms a year for thirty years determined that the amount of nitrosamine absorbed through the skin, 0.9 micrograms in total, is one million times less than the dose that causes tumors in animals.

A more realistic concern involves the degradation of rubber exposed to ozone. Photocopying machines generate a fair bit of this gas, making the storage of condoms near them an unsafe practice. It seems that there are people out there who need to heed this advice. A survey in Britain showed that 28 percent of working women have had sex in the office. (Strangely, men were not surveyed.) Researchers were able to determine that 12 percent of these activities took place on the boss's desk, but the frequency of activities that involved a photocopier is unknown.

TAMPONS ON A MISSION

Physicist Sally Ride blasted into space aboard the Space Shuttle Challenger in 1983 for a mission that was to last eight days, so the possibility of "that time of the month" had to be considered. Since no American woman had flown in space before, NASA had no information about how microgravity might affect menstruation. Two female Soviet cosmonauts had flown in space before Ride, one for three days and the other for eleven, so the Soviets must have addressed the question. However, due to the secrecy surrounding the space race, whatever knowledge they had gathered was not forthcoming.

NASA physicians had some concern that without the influence of gravity blood may flow in the wrong direction and wanted to make sure that it would be absorbed before that could happen. They asked Ride how many tampons she may need, and building in a huge safety factor, came up with the preposterous number of one hundred. Whether Ride had to use the tampons was never revealed, but given the attention being paid to the space program, the episode did generate lots of discussion about tampons. That was useful at the time because women were concerned about Toxic Shock Syndrome (TSS), a bacterial infection that had been linked with tampons.

While modern tampons date back only to the 1930s, attempts to absorb menstrual blood in such a fashion have a long history. Roman women fashioned devices out of wool, and in parts of Africa rolled-up grass was favored. But it was Colorado physician Earle Cleveland Haas who is credited with introducing the prototype of the current product. His wife was a ballerina and Haas was supposedly motivated by her struggles with the pads available at the time. In 1933, he patented a pack of compressed cotton that could be inserted by means of telescoping paper tubes. He called it Tampax from "tampon" and "pack."

Tampons made it into the headlines just prior to Sally Ride's mission when Procter and Gamble's super absorbent Rely, formulated with carboxymethylcellulose and compressed beads of polyester, was linked with TSS. It turned out that the material retained bacterial toxins that would normally pass out of the body with menstrual flow. Other super absorbent tampons also were found to increase the risk of TSS, although not as much as Rely. Current tampons are not made with carboxymethylcellulose or polyester, but that doesn't mean they have not been accused of undermining women's health.

One absurd letter circulating on the web claims that manufacturers add asbestos to tampons to increase bleeding so they can sell more product. Ridiculous. Then there is the allegation that tampons are contaminated with dioxin, a notoriously toxic substance linked with cancer and hormonal disruption. The usual suspect for the source of the dioxin is rayon, used for its absorptive property. Rayon is a cellulosic fiber made from wood pulp and its production involves bleaching. While this produces a white product, the real purpose of bleaching is to remove various impurities.

Classically, bleaching was carried out with chlorine gas, and that turned out to be a problem. Reaction of compounds in wood pulp with chlorine results in the formation of dioxins, nasty substances indeed. But a switch to chlorine dioxide essentially eliminated the formation of dioxins and, according to the FDA and Health Canada, the trace amounts that may be detected are of no consequence. It is also possible to use hydrogen peroxide as a bleach with no chance of forming dioxins at all. Nevertheless, some brands hype "no rayon," which amounts to a sales gimmick. These tampons are usually made with unbleached organic cotton which has no possible contamination with dioxin, but they are also less absorbent.

While there is a lot of misinformation about tampons, there is a real issue that merits discussion. And that is their effect on the environment. The concern is about the plastic applicators that are used to insert tampons. Close to 90 percent of products sold are equipped with plastic instead of cardboard applicators. Considering that a woman can use up to 17,000 tampons in a lifetime, calculations show that some eight billion tampons are disposed of every year in North America. That's a lot of plastic! What happens to it?

Despite clear instructions that the applicators are not to be flushed down the toilet, many of them are. They end up in sewage and find their way into rivers, lakes, and oceans. Some 3,000 were collected in a quick cleanup of seventy beaches in New Jersey in 2014! Aside from the yuck factor, there is a more significant problem. Although the plastic is not biodegradable, it does break into smaller pieces that can be ingested by wildlife. Oysters have been found to be contaminated with microscopic particles of plastic and digestive problems have been detected in fish.

Plastic applicators are not a requirement for tampons, but they do make for greater ease and comfort than cardboard versions. In Europe, tampons sans applicator are more popular because women do not want foreign substances inside them. Although manufacturers are secretive about the exact composition of the plastic, it is likely to be low-density polyethylene, a substance that does not leach any phthalates or bisphenol A, two plastic chemicals that have drawn a lot of attention because of their endocrine disruptive properties. Scented tampons, essentially a marketing gimmick, may, however, contain phthalates.

In America, women are apparently more averse to self-touching and don't like digital application of tampons even though that is the more environmentally friendly way to go about the business. Even more environmentally friendly is the silicone menstrual cup. It is inserted like a tampon, emptied out as needed, cleaned with soapy water, and reused. Economical, no dioxin concern, and minimal risk of Toxic Shock Syndrome. NASA has not disclosed whether any of the forty-four American female astronauts who followed in the footsteps of Sally Ride have tried it.

SAVING *APOLLO 13*

"Tranquility base here, the Eagle has landed." Those words uttered on July 20, 1969, by *Apollo 11* astronaut Neil Armstrong punctuated what may well be regarded as humanity's greatest scientific accomplishment. Just nine years earlier

President Kennedy had initiated the project to "put a man on the moon and return him safely to the Earth before the end of the decade." At the time, scientists did not know how they would accomplish this, certainly not within the time frame of a few years. But ingenuity raced to the forefront, and before the decade was out, the bug-like lunar lander nicknamed the Eagle successfully descended to the moon's surface. A few hours later, Commander Armstrong made the first foray onto a heavenly body other than the Earth, uttering the famous phrase "one small step for man, one giant leap for mankind."

Just four months after the Eagle's historic landing, *Intrepid*, *Apollo 12*'s lunar lander, made a perfect landing on the moon. It was expected that in April of 1970, *Aquarius*, *Apollo 13*'s lander, would do the same at a site designated as "Fra Mauro." But we were never to hear the words, "Frau Mauro here, *Aquarius* has landed." Instead, the world was shaken by astronaut Jack Swigert's iconic words: "Okay, Houston, we've had a problem here." Indeed, there was a problem. On the way to the moon, an oxygen tank in the service module upon which the proper functioning of the command module depended, exploded! The oxygen was needed to combine with hydrogen in fuel cells to generate the electric power, as well as to produce water for the command module. Without the fuel cells, the command module had to rely on a small amount of stored water and on batteries for electricity. The moon landing of course had to be aborted, and plans were quickly made to swing the spaceship around the moon and place it on a path towards the Earth. The battery power in the command module would have to be saved for the maneuvering needed for reentry into the Earth's atmosphere.

That journey would take days, far too long for the astronauts to be able to stay in an incapacitated command module. Plans were drawn up for a transfer to the lunar lander, which would serve as a lifeboat. The *Aquarius* was equipped with an oxygen supply, battery power, as well as food and water that were to have served for the time spent on the lunar surface. But it was not equipped with enough carbon dioxide scrubbers to last the time it would take to make the journey back to Earth.

Humans have to inhale oxygen in order to allow cells to combust carbohydrates and fats for the production of energy. But like any other combustion of organic matter, this results in the release of carbon dioxide. While the gas is not toxic, it is heavier than air and can displace it, meaning that a significant buildup of carbon dioxide in a closed environment can lead to asphyxiation. Such a buildup can happen when the gas is formed by underground volcanic activity or by the decomposition of limestone (calcium carbonate) and then seeps into sheltered places such as caves where it can cause animals to suffocate.

A classic example is a cave beside the Temple of Apollo in Pamukkale, Turkey. The ancient Greeks believed this was the entrance to the underworld since no animal or man who wandered into the misty cave ever returned. Today we have a good idea as to why. The area around the cave is permeated with subterranean hot streams. As the hot water flows over deposits of limestone, it causes it to decompose and liberate carbon dioxide. The pressure of the gas builds up, causing some of it to dissolve in the water. Sort of a natural carbonation process. Then as the carbonated water reaches the cave, the pressure is released and the gas escapes. Kind of like opening a bottle of pop. The carbon dioxide then pushes the air out of the cave, and anyone entering can quickly be overcome by a lack of oxygen.

Obviously then, carbon dioxide has to be removed from the air in sealed systems such as submarines or spaceships. The usual method is to circulate the air through carbon dioxide scrubbers that are filled with lithium hydroxide, a chemical that reacts with carbon dioxide gas to produce solid lithium carbonate. Engineers on the ground went to work and devised an ingenious way to use the lithium hydroxide canisters from the command module that were not compatible in shape to fit into the lunar module's scrubbing system. A clever use of the available canisters, cardboard, plastic bags, and duct tape resulted in a makeshift filter system that saved the astronauts' lives as far as breathing was concerned. But there was still the problem of jettisoning the crippled service module and the lunar lander after the astronauts had climbed back into the underpowered command module. Scientists on the ground, including some at the University of Toronto, devised emergency procedures to do this and then waited with bated breath for a successful reentry through the atmosphere.

Reentry on a lunar mission is accompanied by a communication blackout as the tremendous heat generated by the friction of the capsule traveling through the atmosphere ionizes the air around it and prevents radio signal transmission. This normally lasts about four minutes, but the blackout during *Apollo 13*'s reentry lasted six minutes, about eighty-seven seconds longer than had been expected. This caused a lot of concern on the ground for fear that the explosion had damaged the capsule's heat shield, which could have meant incineration for the astronauts. There was great relief when radio contact was reestablished, followed by a safe splashdown in the South Pacific. Chalk one up for human ingenuity and duct tape.

SINGING ABOUT SCIENCE

Mary Poppins, starring Julie Andrews, was a big hit for Disney Studios in 1964. The film was a musical version of the children's books about a magical English nanny written between 1934 and 1988 by P. L. Travers and featured a number of songs written by Robert and Richard Sherman, including the catchy tune sung by Andrews, "A Spoonful of Sugar." The idea for the lyrics came from a real life situation. Robert Sherman was working on ideas for a song but was drawing a blank until one day he came home and learned from his wife that his children had gotten a polio vaccine. Thinking that the vaccine had been a shot in the arm, he asked one of his children whether it had hurt. Not at all, the child replied. There had been no jab. A drop of liquid was placed on a sugar cube that had to be swallowed. At that moment the title for the song was born!

The oral vaccine that the Sherman children received had been developed by Albert Sabin and was introduced commercially in 1961. It used a weakened form of the poliovirus that triggered the production of antibodies against the active virus. The oral version to a large extent replaced the original injectable vaccine introduced in 1955 by Jonas Salk based on an inactivated form of the virus. Thanks to these vaccines, polio has been largely eliminated from the world. Of course, every sort of medical intervention is associated with some risk. In very rare cases, the vaccine can cause polio symptoms, but the benefits greatly outweigh any risk. Both vaccines are on the World Health Organization's List of Essential Medicines, which identifies the most important medications needed in a basic health system. Had the vaccine been available earlier, President Roosevelt would not have contracted polio in 1921.

The spoonful of sugar in combination with a medicine may have an impact other than just pleasing our musical appetite. It seems that infants given a little bit of a sugar solution feel less pain during injections. British pediatrician Paul Heaton found that a few drops of sucrose solution put on their tongues before an injection was capable of blocking the pain felt in their arms or bottoms. He theorizes that "the sweet taste works through nerve channels in the tongue that perceive sweetness in the brain." The brain reacts by producing endorphins, the body's natural pain relievers. Furthermore, in babies, sucking releases endocannabinoids that also alleviate pain. Heaton noted that once babies taste the solution, they cry less and recover more quickly from the jab. He recommends giving babies just enough sugary solution to taste but not enough to swallow before vaccination. Interestingly, the relationship between sweets and pain relief was first mentioned in historic Jewish texts that document baby boys being given honey before circumcision. What about adults? Well, chocolates, sweet pastries, and soft drinks make for a less painful life for many people.

The Sherman brothers also composed the song that has been played more often in the world than any other. "It's a Small World (After All)," an adaptation of an attraction introduced at the New York World's Fair in 1964, is featured at all the Disney theme parks. The Sherman Brothers wrote the song in the wake of the 1962 Cuban Missile Crisis, which influenced the song's message of peace and brotherhood. They also wrote a song for the Adventure Thru Inner Space attraction that was presented in Disneyland's Tomorrowland from 1967 to 1985 and designed to simulate humans shrinking to a size smaller than an atom. Visitors boarded "Atommobiles" and began a journey that passed through snowflakes into the inner space of molecules, then atoms. They got an idea of crystal structure, bonding between atoms, and the composition of an atom. The journey was accompanied by the song "Miracles from Molecules."

From the beginning until 1977 Adventure Thru Inner Space was sponsored by the Monsanto Company, which later transitioned from being a chemical manufacturer to a biotechnology firm. Founded in 1901 by John Francis Queeny and named after his wife's family, Monsanto initially produced food additives such as saccharin and vanillin before expanding into industrial chemicals such as sulfuric acid and PCBs in the 1920s. By the 1940s it was a major producer of plastics, including polystyrene, as well as a variety of synthetic fibers. Monsanto scientists had a number of notable achievements such as the development of catalytic asymmetric hydrogenation, which made possible the production of L-Dopa, the major drug used in the treatment of Parkinson's disease. They also laid the foundation for the mass production of the light emitting diodes (LEDs) that have revolutionized the lighting industry.

Monsanto has been criticized for once manufacturing controversial products such as the insecticide DDT, PCBs used as insulators in electronic equipment, and the notorious Agent Orange that was widely deployed as a defoliant during the Vietnam War. At the time DDT and PCBs solved immediate problems, with DDT saving millions of people from contracting malaria and PCBs in transformers making electricity widely available. The environmental issues that eventually emerged concerning these chemicals were not, and probably could not have been, foreseen at the time.

Today, most people associate Monsanto with genetic modification. The company serves as a lightning rod for anti-GMO activists. Indeed Monsanto was among the first to genetically modify a plant cell and to conduct field trials of genetically modified crops and now markets canola, soy, corn, and sugar beet seeds that yield plants capable of resisting herbicides and warding off insects.

Let me end with a stanza from the Sherman brothers' song "Miracles from Molecules" that once captivated visitors to Disneyland and which I believe is still meaningful today:

Now Men with dreams are furthering,
What Nature first began,
Making modern miracles,
From molecules, for Man.

THE SKINNY ON SKIN SCIENCE

"All the carnall beauty of my wife is but skin deep," wrote Thomas Overbury in his 1613 poem, "A Wife," with the message being that attractiveness had no relation to inner beauty. Curiously, Overbury had no wife. He penned the poem to express what one should look for in a wife with his friend Robert Carr in mind. Carr had begun an affair with the married Countess of Essex, an affair Overbury believed would have a disastrous outcome. She was already "noted for her injury and immodesty," he wrote. Actually the outcome was indeed disastrous, for Overbury.

The publication of the poem made an enemy of Lady Essex, who believed that Overbury sought to open Carr's eyes to her defects. In order to have the meddlesome man out of the way, she contrived to have King James I offer him an ambassadorship. However, Overbury refused the appointment, desiring to remain in England by his friend Robert's side. An infuriated king responded by having him imprisoned in the Tower of London, where he died. Suspicion was that he had been poisoned with copper vitriol, the plot engineered by the beautiful Lady Essex, who turned out to be quite ugly on the inside.

Ever since that episode, the meme "beauty is only skin deep" has been often repeated to suggest that a person's character is more important than their outward appearance. But that hasn't stopped ladies from doing everything possible to improve their looks. And cosmetic companies are keen to get engaged in this task with the message that beautifying the skin brings happiness. It certainly does to them, as they crank out thousands of profitable products with claims galore.

There are creams that claim to rejuvenate, regenerate, strengthen, energize, nourish, tone, heal, soothe, cleanse, pamper, tighten, repair, buff, brighten, exfoliate, detoxify, improve microcirculation, boost cellular energy, neutralize free radicals, reduce inflammation, regulate cellular proliferation, reduce wrinkles, improve elasticity, stimulate collagen formation, boost skin protein synthesis, accelerate skin recovery, and moisturize. The promise is of magic. But magic is only delivered on the stage, not in a cream. Most of the terms amount to no more than advertising hype, but that doesn't mean the products are useless. They can moisturize effectively and in some cases improve skin structure.

There is no question that the appearance of skin is improved with sufficient moisture content. The basic problem is that skin dehydrates readily, especially if the air is dry. Moisturizers prevent this by coating the skin with a fatty layer that prevents water from evaporating. Petroleum jelly, lanolin, and a variety of oils are very effective moisture barriers and can make the skin softer and smoother. But they feel greasy. The greasiness can be reduced by blending with water, but the problem is that oil and water don't mix. This is where emulsifiers, molecules that have one end that is water-soluble and another that is oil-soluble, come to the rescue by forming a link between oil and water, preventing separation. Hundreds of emulsifiers exist with lecithin, sorbitan stearate, cetearyl alcohol, and polysorbates being typical examples. The challenge of producing a moisturizing cream with the desired consistency revolves around finding the right emulsifier.

When it comes to improving skin structure, the leading ingredients are peptides, molecules composed of chains of amino acids. One of the main components of skin is collagen, a protein that can be likened to a scaffolding that is constantly being formed and broken down through the action of enzymes. When breakdown exceeds formation, the skin sags and wrinkles form. A variety of oligopeptides, short chains of amino acids that are based on sequences actually found in collagen, have been synthesized for incorporation into creams. The idea is to trick the skin into believing that too much collagen has been broken down, causing a reduction in the activity of collagenase, the enzyme that breaks collagen down, and boosting the activity of fibroblasts, cells that produce collagen.

Oligopeptides can reduce fine lines and improve skin smoothness, but not as effectively as injections of botulin, the notorious poison produced by the *Clostridium botulinum* bacterium. Botulin improves appearance by paralyzing the muscles that cause the skin to wrinkle. But the injections have to be given by a physician, and they are expensive. This has spurred research into finding some sort of ingredient that can be incorporated into creams to mimic the effect of Botox. One of the most curious ones is based on peptides found in the venom of the temple pit viper, so named because the snake is found in Malaysia around the Temple of the Azure Cloud. The venom contains a number of peptides that interfere with the action of the neurotransmitter acetylcholine, resulting in the paralysis of its prey.

Including snake venom in a cream is not an attractive marketing feature, so the Swiss company Pentapharm has come up with an alternative idea. Its chemists have analyzed the snake venom peptides and have determined that one particular sequence of three amino acids is responsible for producing the paralysis. This tripeptide has been synthesized, dubbed Syn-ake, and has been incorporated into creams. Does it work? The company has produced a study involving forty-five volunteers who used either a cream with Syn-ake, a placebo cream, or one with acetylhexapeptide-3, a competing anti-wrinkle peptide that supposedly mimics to some extent the action of botulin. Syn-ake performed better than either, with an improvement in skin roughness and a decrease in the depth of wrinkles. However, the measurements were made by using sophisticated instruments; it is unclear whether an objective observer would note a difference in the subjects' appearance. The trial has yet to be published in the peer-reviewed literature. Still, there is some interesting science there, so oligopeptides in creams cannot be passed off as snake oil.

SLIMY SCIENCE

I love to wander through cosmetic trade shows. Lots of entertaining hype with truckloads of hope squeezed into pretty jars. But you can also find some intriguing science. I stopped at a booth featuring "snail cream." No, you don't eat it. Neither is it meant to limber up the arthritic joints of snails. You massage it into your face to "improve complexion, reduce wrinkles, and improve scar lines." Creams that contain snail slime are a hot item in South America and Korea and are slowly slithering their way to North America. Sounds like another slimy marketing effort, but don't roll your eyes quite yet. It seems Hippocrates favored a mix of sour milk and crushed snails for inflamed skin. Of course that doesn't mean it worked. After all, the man who gave us the Hippocratic Oath also thought that pigeon droppings cured baldness. But there may be something to the snail gambit. Apparently Chilean farmers who were raising snails for the food market noted that their skin became smoother after handling the creatures. Not exactly scientific evidence, but enough for the cosmetic industry to pick up some speed in marketing snail slime.

The slime is a complex chemical mixture that contains proteoglycans, glycosaminoglycans, and a variety of enzymes. There's also hyaluronic acid, some copper peptides, and antimicrobial compounds. All these combine to protect the snail from cuts, abrasions, and bacteria. But what can they do for people? There is actually evidence that at least some slime components can stimulate the proliferation of fibroblasts, the cells that produce collagen and elastin, the proteins that form the basic matrix of skin structure. The problem, though, is that these effects have only been seen in cell culture. There's no study that has documented a benefit from snail cream in people. Even without such evidence, some producers claim to have a superior product because their snails are raised on ginseng!

And how does one get snail slime? It seems the snails have to be stressed to secrete the sticky stuff. I'm not sure how one stresses a snail — perhaps by signing it up for a race, or by frightening it with one of those dishes with the six little wells ready to be loaded with escargots. Eventually we may actually see some scientific evidence for the benefits of snail cream, but while the hype races ahead at breakneck speed, the research seems to progress at . . . well . . . a snail's pace. The cream is expensive, but if you want to give snail slime a try, maybe you can find some snails in the garden that will crawl over your face for free. And if snail cream isn't exotic enough, I also came across shampoo with bull semen, which apparently "leaves hair with a brilliant sheen that no other substance can match." Very pricey, but I suppose collecting the needed ingredient presents some occupational hazards.

The terms most often featured at trade show displays these days are "natural," "organic," and "sustainable." The drive towards safe, environmentally friendly products is admirable, and the "green" philosophy of using chemicals that have maximum safety and minimal impact on the environment certainly should be followed. But given that there are no universal standards on what constitutes "natural," "organic," or "sustainable," the terms are often flung about in a reckless fashion.

If you squeeze the flesh of a coconut and collect the oil to use in a cosmetic, everyone would agree that you have a natural, organic ingredient from a sustainable source. We'll ignore for now the ecological consequences of burning down rain forests in Asia in order to establish palm tree plantations. But what happens if you now extract lauryl alcohol from the coconut oil? That would still be natural because it was originally present in the oil. And what if now you treat this oil with sulfuric acid and then with sodium carbonate to produce sodium lauryl sulfate, a surfactant present in a variety of cosmetic products? Sulfuric acid can be made by burning sulfur, a naturally occurring substance, and sodium carbonate also can be found in nature. Some would therefore say that sodium lauryl sulfate is natural, others would argue that it is not because chemical reactions have been used to produce it. Chemical reactions are fine for some, as long as no petroleum-derived substances are used. A curious view, since petroleum is the decomposition product of plants and animals and can therefore also be regarded as "organic."

Perhaps the most effective marketing claim these days is "scientifically proven to be effective." But this can have different meanings. For example, there are instruments that create a vacuum on a small patch of skin and measure how fast the skin regains its shape when the suction is released. Another, a "ballistometer," drops a ball on the skin and measures how the skin rebounds after the impact. Such tests are indeed scientific, and can show an improvement after a product is applied to the skin, but this does not necessarily mean that there will be an observable difference on looking into a mirror. There are also various imaging instruments and magnifiers that measure skin pigmentation and wrinkle depth and can be used to demonstrate a "before and after" difference, but again, this may not translate to ooohs and ahhhs from admirers.

On further wandering through the booths, I came across extracts of licorice, sage, saffron, sea fennel, horse chestnut, tea, and turmeric, all hyping their antioxidant, antiaging, and "bioprotectant effects." A "proprietary blend of youth-enhancing ingredients extracted from fermented rice" promised to deliver a "complete remodeling effect by increasing skin firmness and erasing deep wrinkles." The only evidence provided was that Japanese brewers who handle fermented rice while making sake have astonishingly smooth, young-looking hands. Then there was a "rejuvenating" cream featuring extract of pomegranate flower. A handout declared that, according to legend, the pomegranate was particularly appreciated by mermaids. Pretty fishy marketing. But some will go for it hook, line, and sinker.

EASTER ISLAND MAY PROVIDE CLUES TO AGING

Easter Island in the South Pacific is famous for the Moai, giant human figures with disproportionately large heads carved out of stone some five to eight centuries ago. The statues came into the public spotlight in 1968 with the publication of Erich von Daniken's inane book, *Chariots of the Gods*, in which he suggested that the technology needed to carve the figures and transport them around the island was imparted to the natives by extraterrestrial visitors. The statues, he claimed, were dedicated to these visitors who were regarded as gods. Balderdash! While the production and transport of the statues was a remarkable creative and physical feat, it did not require alien intervention. The figures are actually believed to be representations of important tribal leaders who had passed on. Recently, the spotlight has again focused on Easter Island, but this time not on the mysterious figures. It is the soil upon which the ancient Moai stand that may hold a key to solving the mystery of aging.

Growing old is not a particularly enjoyable experience. Cleopatra tried to counter aging by taking baths in donkey milk. The alchemists believed that the secret lay in finding a way to formulate gold, a metal that never seemed to age. In the sixteenth century, Hungarian countess Elizabeth Bathory sought to prevent the ravage of passing years by bathing in the blood of female virgins who were murdered after being hired as maidservants by the bloodthirsty countess. Mary, Queen of Scots, resorted to white wine baths to preserve her youth, and in the nineteenth century, testicle transplants from goats or monkeys into the scrota of aging men were all the rage. The twentieth century introduced hormone replacement therapy, human growth hormone injections, Botox, Viagra, and a host of tantalizing "youth in a bottle" supplements ranging from various antioxidants to fish oil and deuterated fatty acids. But the only regimen that has shown any sign of success has been calorie restriction. Yeasts, worms, flies, and rodents all live longer with a reduced caloric intake! This has not been demonstrated in humans, but in any case, cutting back on calories in an extreme fashion is not an appealing prospect. Popping an antiaging pill is a far more attractive proposition than starvation. And that brings us to the soil of Easter Island.

It was back in 1964 that Canadian scientists became interested in why the island's natives were not affected by tetanus despite their barefoot culture. Curiously, as it turned out, spores of the tetanus bacterium were rare in the soil samples they collected. Eventually these samples were donated to the pharmaceutical company Ayerst, which was interested in soil microbes with potential biological activity. It was there that in 1972 Dr. Suren Sehgal isolated a bacterium, *Streptomyces hygroscopicus*, that produced a compound with antifungal activity. Sehgal named it "rapamycin" after Rapa Nui, the native name for Easter Island. Further research revealed that besides the antifungal activity, rapamycin had a powerful immunosuppressant effect, leading to its introduction in 1999 as Rapamune for the prevention of rejection of transplanted organs.

What really excited researchers, though, was the finding that rapamycin also stopped cancer cells from dividing. And it did so without killing the cells! That was an exciting development because most cancer drugs also affect healthy cells, resulting in the classic side effects of chemotherapy. There was, however, a problem. The original patent for rapamycin did not include its use against cancer, which meant that any company could produce it for this purpose, reducing the potential for financial profit. In such cases, the approach is to slightly alter the molecular structure of the compound in question to make it a novel, patentable substance. But carrying out a molecular modification such that biological activity is retained requires insight into how a drug actually works.

Researchers at Wyeth, the pharmaceutical company with which Ayerst had merged in 1987, went to work on this problem and found that rapamycin binds to a specific protein for which they, unimaginatively, coined the term "mechanistic target of rapamycin," or mTOR. This protein stimulates cell growth, kicking into action based on the levels of insulin, glucose, leptin, amino acids, and oxygen, the nutrients that cells need to grow. Blocking the action of mTOR is then expected to prevent cell growth and multiplication, including that of cells critical to the functioning of the immune system. Elucidation of the mechanism by which rapamycin binds to its target opened the way to patentable derivatives, referred to as rapalogs. Temsirolimus (Torisel) and everolimus (Afinitor) eventually were introduced for the treatment of kidney cancer.

But how does all this connect to aging? When cells are processing nutrients and growing, they also produce side products. These are the notorious free radicals that have been linked with aging on account of their ability to destroy nucleic acids and proteins. Reducing the activity of mTOR reduces cell growth and thereby reduces the damage caused by free radicals. Calorie restriction accomplishes this by causing mTOR to gear down due to lack of nutrients. But now it seems that rapamycin and its analogs may do the same without us having to starve ourselves. There is, however, the niggling issue of immune suppression. The question is whether the right dose of rapalogs can reduce mTOR activity without impairing immune activity.

In 2014, Novartis reported a trial in which 200 elderly people took either placebo or everolimus over the six weeks prior to receiving a flu vaccine. At just the right dose of the drug, antibodies were significantly increased, meaning that the immune response was modulated, not suppressed. This means that the antiaging effect of rapalogs can be further explored without concern about increasing the risk of infection. Those aging statues on Easter Island may be standing on a version of the much sought-after fountain of youth. But a lot more research is needed before we drink of it.

A TALE OF TELOMERES

As you get older, your telomeres get shorter. That much is certain. But whether you can lengthen life by lengthening your telomeres is a different question. Wrinkles are also a sign of aging, and while a facelift may stretch your skin, it won't stretch your years. So are shortened telomeres a result of aging or a cause of it?

Promoters of a supplement derived from the *Astragalus membranaceus* herb believe they know the answer. If you want to maintain health and lengthen your life, you need to lengthen your telomeres! You also need to open your wallet. Wide. "Telomerase activator-65 (TA-65)" will set you back about $4,000 for a six-month supply! But for that, at least according to the ads, you are getting "Nobel Prize Technology."

Well, that's not exactly so. The Nobel Prize in Physiology or Medicine in 2009 was awarded to Elizabeth Blackburn, Carol Greider, and Jack Szostak for their work on telomeres. That work, though, had nothing to do with the dietary supplement being promoted. It had to do with explaining how short bits of DNA known as telomeres protect the long strands of DNA packed into chromosomes from degradation during cell division. A common analogy for telomeres is the plastic protective cap on the end of a shoestring. Degradation of chromosomes is associated with the aging of cells, and that in turn is linked with aging of the whole organism.

When cells divide, chromosomes are copied, and if not properly protected by the telomere caps, they become damaged. With each division, though, the telomeres become a little shorter, possibly an ominous situation. Back in 2003, researchers at the University of Utah analyzed blood samples from 143 elderly people and found that those with the shortest telomeres died on average four or five years earlier than others. The implication was that perhaps a simple blood test could foretell longevity! Numerous research groups jumped on the bandwagon and by 2013 amassed a wealth of conflicting data. Some found an association between telomere length and early death, others did not. Some concluded that stress shortened telomeres, others that less time spent sitting lengthened them. A Mediterranean diet seemed to protect telomeres, guzzling soft drinks damaged them.

The soft drink study sparked a media frenzy with some accounts claiming that drinking soda has the same effect on aging as smoking cigarettes, possibly shortening life by half a decade. The study got a lot of traction given that one of the authors was Elizabeth Blackburn, the Nobel Laureate. A careful look at the research, however, raises questions. The data was based on blood samples taken from some 5,000 people who were interviewed in a national survey about their food and beverage consumption during the past twenty-four hours. Most of us can't remember exactly what we ate for breakfast, never mind giving a full and accurate description of what we consumed yesterday. In any case, a first appraisal of the data revealed no link between telomere length and soft drink consumption.

Further data dredging did, however, unveil an association between carbonated soft drinks, but not other beverages, and shorter telomeres. Based on the cherry-picked studies that had found a connection between reduced longevity and shorter telomeres, the researchers concluded that people who drank two soft drinks a day shortened their lives by about four and a half years. Elissa Epel, one of the authors, opined that "the extremely high dose of sugar that we can put into our body within seconds by drinking sugared beverages is uniquely toxic to metabolism." Indeed, the damage was said to be equivalent to that caused by smoking a pack a day.

Really? How is it then that energy drinks and fruit beverages that contain the same amount of sugar as bubbly drinks were not associated with shortened telomeres? Would the proper conclusion not have been that the problem is carbonation? I do consider sugared beverages to be a nutritional curse, but this study does not prove that they shorten life, and equating drinking soda pop with smoking suggests that there may be some other agenda here. Like promoting the idea that "cellular aging" can be determined by telomere length and perhaps slowed by various lifestyle interventions.

Epel has suggested that meditation, yogic chanting, and various relaxation techniques owe their benefits to their effect on telomeres. She has expressed these sentiments at Deepak Chopra's Sages and Scientists Symposium, which is a get-together of individuals who routinely pay homage to Chopra's confused ideas about nebulous concepts such as "quantum healing." Can a book about simple methods to lengthen telomeres and life be in the offing? And an appearance on *The Dr. Oz Show*?

Although the evidence linking telomere length to diet, exercise, and meditation may be pretty thin, there is plenty of other evidence to support the benefits of these lifestyle factors. But the story is different when it comes to dietary supplements that claim to boost the activity of telomerase, the enzyme that builds telomeres. If you are going to fork out a small fortune, you should want to see some powerful evidence. Is there any? Hardly.

In one experiment, when fed the TA-65 supplement, geriatric mice showed a lower percentage of very short telomeres. They also had lower insulin levels, increased hair regrowth, and plumper skin, but longevity did not change, neither did average telomere length. Overall, not very impressive. As far as human studies go, there is only one, and it was sponsored by the company. In thirteen older men and women who took the supplement for twelve to eighteen months, the average telomere length did not change, but some clever data mining revealed that in the cells of seven of the volunteers, the percentage of short telomeres declined. No antiaging effects were noted. Despite this, people like supplement peddler Dr. Al Sears have no problem promoting and selling TA-65, declaring it to be the "most important discovery in human history." Sears also believes that "Big Pharma" killed President Eisenhower with its misguided attempts to lower his cholesterol, which he thinks has nothing to do with heart disease.

By all means, exercise, meditate, and eat your fruits, veggies, whole grains, and nuts. Drizzle olive oil on your salad to your heart's delight. Shun sugary soft drinks. But leave the connection to telomere length for researchers to figure out. Maybe someday there will be enough evidence to recommend some supplement that lengthens telomeres. That day is not now. Furthermore, there is the somewhat worrying observation that cancer cells multiply quickly, possibly because they maintain their telomere length. Finally, ponder this: If TA-65 really did what it claims to do, namely affect a fundamental change in cells, shouldn't it be regulated as a drug?

SCIENCE SNIFFS AT BODY ODOR

"The odors of wood, iron, paint and drugs cling to the garments of those that work in them. Thus I can distinguish the carpenter from the ironworker, the artist from the mason or the chemist. Human odors are as varied and capable of recognition as hands and faces." Those words belong to Helen Keller (1880–1968), who became a famous writer, lecturer, and political activist despite being deaf and blind. She was not born with these afflictions. They were the result of her contracting what is believed to have been either scarlet fever or meningitis when she was nineteen months old.

Eventually through the tutelage of Anne Sullivan, Helen learned to read braille and speak, becoming the first ever deaf and blind person to earn a Bachelor of Arts degree. As often happens when one or more of the senses are impaired, the others become more acute. Helen Keller developed a keen sense of smell, such that she was able to identify roses and fungi by their scent, even being able to identify the deadly *Amanita muscaria* mushroom by its fragrance.

Smell is the brain's interpretation of volatile compounds interacting with receptors on cells in our nasal passage. These receptors are protein molecules that are twisted into various shapes, forming pockets into which volatile molecules fit, much like a hand fits into a glove. There are numerous such receptors that have evolved to accommodate the wealth of volatile compounds to which we can be exposed. The scent of roses, for example, is due to a blend of dozens of compounds, among them citronellol, geraniol, phenyl ethyl alcohol, limonene, rose oxide, alpha-damascenone, beta-damascenone, benzaldehyde, benzyl alcohol, and phenyl ethyl formate. Different varieties of roses will produce specific, identifiable arrays. Some of these compounds also occur as components of other smells. Benzaldehyde, for example, is part of the fragrance spectrum of apricots, apples, and oyster mushrooms. Beta-damascenone is found in coffee aroma along with over a thousand other compounds!

Similarly, human body odor is an extremely complex mixture of compounds with a virtually infinite possibility of combinations. Where do these compounds come from? There are thousands and thousands of compounds floating through our bloodstream and lymphatic system all the time, originating either directly from food or from the numerous metabolic reactions occurring in the body all the time that together constitute life. Some of these are exuded through the skin and are wafted into the air, ready to engage with nearby nostrils.

Eat some garlic and allyl methyl sulfide will turn up in your breath, pee, and sweat. Some volatile compounds such as steroids such as androstenone are produced in the body from cholesterol, which in turn is synthesized from simple dietary components. This compound has received attention as a putative human pheromone based on the observation that it is found in boar saliva and is capable of triggering mating behavior in sows. Androstenone is the hyped ingredient in many supposed aphrodisiacs with essentially no evidence. An interesting feature, though, is that not everyone can smell androstenone, usually described as a musky or urine-like scent by those who are equipped with the right genetics.

Other smells can be traced to problems with metabolism. A lack of insulin, as in diabetes, causes the body to use fats instead of glucose for energy resulting in fat metabolites such as acetone showing up in the breath and sweat. Trimethylaminuria is a rare genetic disorder in which the body is unable to process trimethylamine, a breakdown product of choline, a common dietary component. Instead of being converted to trimethylamine oxide, which has no odor and is passed through the urine, it is eliminated by passage through the skin, where it produces a disturbing rotten fish odor.

Most compounds that are found in the sweat of healthy people are not volatile enough to produce a scent until they are broken down into simpler molecules by the roughly one thousand varieties of bacteria that inhabit our skin. The composition of this flora is determined both by genetics and environmental exposure. Every square centimeter of our skin is covered with roughly a million bacterial cells that churn out enzymes capable of breaking down the fats, proteins, and vitamins secreted by our sweat glands into smelly, volatile compounds.

In addition to steroids, the pungency of underarm odor is mainly due to compounds that fall into the thiol and carboxylic acid families. Thiols are very nasty smelling compounds, epitomized by the stench of skunk secretions. Carboxylic acids can also be malodorous, such as butyric acid, the smell of rancid fat. A specific thiol (3-methyl-3-sulfanylhexan-1-ol) and a specific acid (3-hydroxy-3-methylhexanoic acid) have been identified as the major components of human sweat malodor, both arising from the action of bacteria on protein metabolites secreted by sweat glands. Interestingly, women liberate more of the fruity and onion-like acid than men, likely due to gender differences in bacterial composition.

Underarm hygiene can control odor by washing away the smelly compounds, deodorants can curb their production by eliminating some bacteria, and antiperspirants can constrict sweat glands and reduce their ability secrete sweat. But there may be a totally novel treatment on the horizon. Molecularly imprinted polymers (MIPs), best described as synthetic antibody mimics, may be making an appearance in cosmetics. Like antibodies, MIPs are capable of specific molecular recognition. They are made in the lab by synthesizing a polymer around a template molecule, such as an odor precursor. Subsequent removal of the template leaves a cavity into which a molecule with the same shape as the template will fit. By scavenging the odor precursor, an appropriate MIP will prevent its conversion into a smelly volatile compound by bacterial enzymes without disrupting the microbial equilibrium on the skin that helps to protect the body against colonization by disease-causing organisms. Armpits may breathe a sigh of relief.

PERFUME AND TNT

One of the most fascinating facets of chemistry is the process of discovery. Think of TNT and chances are you think BOOM, not Chanel No. 5. But trinitrotoluene (TNT) played a major role in the formulation of one of the most famous fragrances in the world.

Following William Henry Perkin's 1856 accidental discovery of mauve, the world's first synthetic dye, the chemical industry was hot on the trail of new colorants. It was then that German chemist Joseph Wilbrand synthesized TNT, which never made it as a yellow dye but did announce itself with a bang. At the time, trinitrophenol, commonly known as picric acid, was the most widely used high explosive but was prone to accidental detonation during production and transport. TNT, on the other hand, can be melted and poured into shell or bomb casings with safety. Its detonation requires the use of a more sensitive explosive such as lead azide, which when energetically struck, quickly decomposes to elemental lead and nitrogen gas. The shock wave created by the rapidly expanding nitrogen sets off the TNT. This same chemistry is used in automobile air bags where sodium azide supplies the nitrogen needed to inflate the bag.

When TNT detonates, it also releases nitrogen along with steam and carbon monoxide. It is the rapid production and expansion of these gases that characterizes an explosion. While TNT never made it as a dye for fabrics, during World War I it did manage to taint the skin of munitions workers, most of whom were women. "Canary girls" these ladies came to be called. Skin discoloration, however, wasn't the only problem. TNT can be absorbed through the skin and cause nausea, loss of appetite, and liver problems. Many workers suffered before it was discovered that application of grease to the skin would prevent absorption.

After the explosive potential of TNT was recognized, chemists went to work trying to get more bang for their buck by attempting to modify the compound's molecular structure. And that is just what Albert Bauer was doing in 1888 when he used the well-known Friedel–Crafts alkylation reaction to add a four-carbon fragment known as a tertiary butyl group to the molecule. As the reaction proceeded, he noted that the lab filled with a decidedly unusual smell. Being a chemist, Bauer was familiar with all sorts of odors, and this one reminded him of the fragrance of musk. That was an exciting observation because at the time musk scent was a much sought-after commodity, highly prized by the perfume industry. Not only did it lend a pleasing note to a perfume, but it also acted as a fixative, slowing down the evaporation of all the perfume's components.

Musk scent was very expensive because of the scarcity of its source, the sex glands of the Asian male musk deer. The animal secretes a smelly mixture of compounds from the glands located near its anus to attract the female. In its concentrated form, the scent is decidedly unattractive, but it becomes seductive when diluted. How anyone ever discovered that the dried and then diluted secretions from this inconspicuous little abdominal sac of the male musk deer charmed receptors in our nasal passage remains a mystery. What we do know, however, is that musk fragrance has been used by perfumers since antiquity, with the word "musk" itself deriving from the Sanskrit word for testicle. The ancient Hindus seemingly were better at perfume-making than anatomy since the scent glands are quite distinct from the animal's testes.

As soon as Bauer sniffed the musky aroma of his new compound, he recognized that he was on to something. He quickly filed a patent for Musk Bauer, and proceeded to make other "nitro musks" with even more effective scents. Musk xylene, musk ketone, and musk ambrette revolutionized the perfume industry and made Bauer a rich man. The nitro musks became the cornerstone of the perfume industry, accounting for the popularity of perfumes such as Chanel No. 5, introduced in 1921, and L'Air du Temps in 1948. They were mainstays until the 1980s when they were dropped because of concerns about their poor biodegradability, neurotoxicity, and tendency to cause a skin rash when exposed to sunlight.

Marilyn Monroe apparently wasn't worried about exposure to sunlight in her bedroom. When asked what she wears to bed at night, she famously replied, "Why Chanel No. 5 of course!" Quite an explosive remark in those days.

GOAT STENCH

Think of a ghastly smell. Skunk? Halitosis? BO? Outhouse? Rotting fish? Rancid butter? Dog flatus? Decomposing flesh? All devastating. But let's not forget the penetrating fragrance of a billy goat. Especially a wet one. That will horrify any nose. Unless that billy goat, or buck in more scientific terms, happens to be castrated. Along with the loss of manhood comes the loss of smell. Actually "smell" doesn't do the aroma justice. "Reek" is a better description of the unforgettable stench. And if you handle one of these animals you will learn what "unforgettable" means. The piercing odor sticks to clothes and skin and is very tough to eliminate. You don't want to be wearing clothes you are fond of when you have an encounter of the male goat kind. And gloves are definitely the order of the day.

Since wethers, as males that have been deprived of their

testes are known, produce no smell, it stands to reason that the aroma of an intact male has a connection to reproduction. That has actually been demonstrated. When exposed to the scent of a male goat, females will ovulate and become receptive to the advances of the buck. Determining the exact composition of the scent that activates the female is of interest to researchers because it may lead to a way of ensuring that females are in rut for breeding. This could be a more economical and a less invasive way of stimulating ovulation than the hormone treatments now used by some breeders. Obviously, the first step in such research is to collect a sample of the scent. Although the identity of the brave scientist who first carried out the pioneering research has been lost to the pages of history, it is known that the scent is wafted out from glands on the billy's head.

So how do you collect the fragrant compounds? You design a helmet equipped with a material that absorbs volatiles, extract these with a solvent, and subject the solution to analysis by gas chromatography and mass spectrometry. A gas chromatograph separates the components of a mixture and a mass spectrometer can identify the individual compounds. And there are lots! Dozens! Determining which are responsible for stimulating ovulation in the female is a challenge, one that was cleverly met by a group of Japanese researchers. By implanting electrodes in a specific area of the brain of female goats, they managed to measure electrical signals associated with the firing of nerve cells that are involved in the release of the hormones that stimulate ovulation. Starting with mixtures of compounds, and then narrowing these down to fewer and fewer components, they eventually managed to determine that 4-ethyloctanal produced the strongest response. It now joins the array of compounds recognized as pheromones, namely, chemicals secreted by animals that influence the behavior or physiology of others of the same species. Interestingly, 4-ethyloctanal has a rather pleasant citrus-like odor.

While this compound has not been found in nature before, it has long been familiar to perfumers and artificial flavor manufacturers. A patent back in 2002 was filed for the use of 4-ethyloctanal as a fragrance chemical to enhance the bouquet of perfumes, toilet waters, colognes, and other personal products, as well as for its use as a food additive to boost flavor. If 4-ethyloctanal isn't responsible for the torturous stench of male goats, what is? The chromatographic analysis reveals a number of compounds in the family of carboxylic acids. These are widespread in nature, the simplest ones being formic acid found in ant venom and acetic acid, a dilute solution that we know of as vinegar.

"Simple" in this case refers to the number of carbon atoms in the acid's molecular structure, with putridity increasing with the number of carbons. Propionic acid, which has three carbons, elicits the scent of body odor, which is not surprising since it is produced by bacteria on the skin. Butyric acid is responsible for the characteristic smell of human vomit. It is also found in goat scent. As disturbing as it is, it pales in comparison with the stabbing odor of the six-carbon caproic acid, appropriately deriving its name from "*caper*," the Latin for goat. Even if you have never been near a goat, you may have encountered its dramatically unpleasant odor if you have ever sniffed the decomposing seeds of the ginkgo biloba tree. Two other acids, caprylic and capric, also derive their names from "caper," because they also contribute to the fragrance of male goats. Staying away from randy old goats is a good idea. Unless they have been deprived of their privates.

PRESERVING PRESERVATIVES

Stories about recalls of various consumer products are all too common these days, but one about contaminated children's sunscreen lotion caught my attention. Not because it posed significant risk, which it didn't, but because the report mentioned "glucono delta lactone." This was a compound I had worked with extensively back in my graduate school days, using it as a starting material for the synthesis of various carbohydrates. What was it now doing in a story about a sunscreen recall?

Cosmetic products, particularly those that are water-based, are prone to contamination by bacteria, molds, and fungi. This is not only a "cosmetic" problem, as it were, it is also a health issue. One would therefore presume that the inclusion of preservatives to ensure a safe product would be seen by consumers as a positive feature, but such is not the case. Preservatives are regarded by many as nasty chemicals that are to be avoided.

This mistrust can be traced back to a 2004 paper by Dr. Philippa Darbre of the University of Reading that described finding traces of parabens, a commonly used class of preservatives, in breast tumors. The study received extensive press coverage with few accounts pointing out that there had been no control group. Since parabens are widely used in foods and cosmetics, they can conceivably be detected in most everyone.

Although Darbre admitted that the presence of parabens did not prove they caused the tumors, she did alarm women by pointing out that these preservatives have estrogen-like activity and that such activity has been linked to breast cancer. What she failed to mention was that the estrogenic activity of the various parabens is thousands of times less than that of estrogenic substances found in foods such as soybeans, flax, alfalfa, and chickpeas, or indeed of the estrogen produced naturally in the body.

Regulatory agencies around the world have essentially dismissed Darbre's study and maintain that there is no evidence linking parabens to cancer. Dr. Darbre, undoubtedly disturbed by being rebuffed, has continued to publish research about parabens, attempting to justify her original insinuation of risk. Her latest paper describes the enhanced migration of human breast cancer cells through a laboratory gel after twenty weeks of exposure to parabens. One is hard pressed to see the relevance of this "in vitro" experiment to the use of 0.8 percent parabens in a topically applied cosmetic.

Nevertheless, because of the concerns that have been raised about parabens and other synthetic preservatives, the cosmetics industry is turning towards the use of "natural" substances that have an unjustified public image of being safer. As I have said many times before, the safety and efficacy of a chemical does not depend on whether it was made by a chemist in a lab or Mother Nature in a bush. Its chemical and biological properties depend on its molecular structure, and the only way to evaluate these is through appropriate experiments.

It is through such experiments that glucono delta lactone's ability to impair the multiplication of microbes was determined. In solution, the compound slowly converts to gluconic acid, creating an inhospitable acidic environment for bacteria and fungi. Marketing-wise, glucono delta lactone can be labeled as "natural" because it can be found in honey and various fruits, where it is formed from glucose by the action of enzymes released from *Aspergillus niger*, a ubiquitous soil fungus that commonly taints plants. Industrially, glucono delta lactone is produced by fermenting glucose derived from corn or rice with the same fungus. But acidification alone is not enough to eliminate the risk of microbial contamination, so the producers of the children's sunscreen turned for help to that spicy mix of vegetables known as kimchi.

Korea's national dish is traditionally made by fermenting cabbage, cucumber, and radishes with the bacterium *Leuconostoc kimchii.* One of the products secreted by the bacteria during the fermentation process is a peptide (a short chain of amino acids) that has antimicrobial properties. Leucidal Liquid is a commercial extract of the antimicrobial peptide produced by the action of *Leuconostoc kimchii* on radishes. In combination with glucono delta lactone it forms an effective preservative system, but as evidenced by the sunscreen recall, not in all cases. The lotions were free of contaminants before being shipped to retailers, but some samples on the shelf were later found to contain bacteria and fungi that could have caused a problem if absorbed through cuts or lesions.

Contamination would most likely not have occurred if parabens, a far more effective preservative, had been used. But the label could then not have declared the product to be "natural." And here we have a curiosity. Compounds in the paraben family actually do occur in nature. Methylparaben can be found in blueberries and, interestingly, in the secretions of the female dog, where it acts as a pheromone notifying the male that its advances are welcome. But since extracting parabens from berries or canine secretions is not commercially viable, the compounds are produced synthetically. This means that even though the final product is identical to one found in nature, it cannot legally be called "natural."

A further issue, at least in the eyes of the chemically unsophisticated, is that benzene, the starting material for the synthesis, is derived from petroleum. Thanks to activist dogma, labeling any chemical these days as "petroleum-based" is tantamount to calling it toxic. So far, no manufacturer has tried to counter this assault by describing petroleum as an organic substance formed through the natural decomposition of biological matter by soil-dwelling microbes, but similar seductive innuendo about "natural" ingredients is not uncommon in the cosmetics industry.

Phenoxyethanol is sometimes advertised as a natural alternative to parabens because it occurs in green tea, but in fact it is commercially made from petroleum-derived phenol. Some companies tout sodium hydroxymethylglycinate as a natural preservative based on the fact that it is made from glycine, an amino acid abundant in the human body. But glycine has to be put through a series of synthetic modifications to produce the preservative.

The demonization of synthetic preservatives has led not only to the glorification of less effective natural products but to a host of "preservative-free" ones as well. These should only be trusted if they come in either single-use vials, or if the sterilized contents are sealed in a container with a pump that prevents entry of microbes when it is used. Otherwise "preservative-free" can quickly become "bacteria-filled."

ANTIBACTERIAL CONCERNS

Store shelves these days sag under the weight of antibacterial soaps, cosmetics, socks, toys, and even garbage bags. There's no question that "antibacterial" on a label increases sales, but there are plenty of questions about the wisdom of impregnating everything in sight with compounds that kill bacteria indiscriminately. Triclosan has been the hot antibacterial in household products for about four decades, but it is now itself feeling the heat due to concern about endocrine disruption, the promotion of antibiotic resistance, and its effects on aquatic ecosystems.

The state of Minnesota has already passed legislation to phase out triclosan except in a medical setting and regulatory agencies around the world are considering doing the same. Companies such as Johnson & Johnson, Avon, and Colgate-Palmolive are all planning to remove triclosan from their formulations. This brings up an immediate question about whether triclosan is to be replaced by some other antibacterial. Quaternary ammonium compounds are likely candidates, but they also come with baggage. Exposure to these compounds has been linked with respiratory irritation, triggering of asthma, and exacerbation of asthma in those who are already asthmatic.

It stands to reason that the use of any chemical should be based on a proper evaluation of risk versus benefit, but such an evaluation is often problematic. Triclosan was first registered as a pesticide in 1969 and quickly found its way into the operating room to replace hexachlorophene as a surgical scrub. It was less toxic, more effective, and more biodegradable, so the benefits greatly outweighed the risks. Triclosan also proved to be useful in protecting adhesives, plastics, caulking compounds, carpets, sealants, and fabrics from attack by bacteria, fungi, and mildew. There is no great issue here because any leaching from these products is minimal. However, the story is different when it comes to soaps, deodorants, shaving creams, cosmetics, dishwashing liquids, and toothpaste, residues of which go down the drain. Here the risk-benefit ratio has been the subject of some bitter controversy.

A fear of bacteria is legitimate, although not of all bacteria. Most live happily in our body and on our skin without causing any harm. But indeed there are the pathogenic varieties that can cause a great deal of misery. *Salmonella*, *Listeria*,

Campylobacter, *Streptococcus*, *E. coli*, *Staphylococcus*, *Clostridium botulinum*, and *Mycobacterium tuberculosis* are worthy of dread, being responsible for hundreds of thousands of cases of illness every year as well as a significant number of deaths.

Aside from toothpaste, where there is actual evidence that 0.3 percent triclosan can help reduce cavities, plaque formation, and gum inflammation, there is no compelling evidence that the addition of triclosan to household products reduces bacterial illness. True, antibacterial soaps can be shown to reduce bacterial counts more than regular soap, but that is not the same as demonstrating a reduction in infections. Marketing seems to have trumped science here.

Another point is that many of the diseases germophobes worry about are caused by viruses, which are unaffected by antibacterials. The viruses that cause the common cold, hepatitis, and many gastro problems scoff at antibacterials. Triclosan may even cause mutations in some viruses, possibly enhancing the risk of viral infection. More importantly, ordinary soap works as well as antibacterial soaps in getting rid of bacteria as long as hands are properly washed, fifteen seconds on each side.

The development of bacteria that are immune to antibiotics is a significant concern. Whether or not bacteria can become resistant to triclosan, and whether triclosan can induce resistance to other antibiotics are hotly debated topics, as is the issue of what happens to all the triclosan that enters the environment from our array of antibacterial consumer products. Wastewater treatment does not eliminate triclosan. About 4 percent is discharged into natural water systems, including those that supply our drinking water, and the rest remains in sewage sludge, which often ends up being used as fertilizer. Here residual triclosan may interfere with the action of bacteria that help fix nitrogen and may even affect earthworms.

Some studies have shown that triclosan can react with the chlorine used to disinfect drinking water to form chloroform, an established carcinogen, and that under the influence of sunlight it can even form small amounts of the notorious dioxins. Then there is the matter of endocrine disruption, with concern being raised about triclosan's chemical similarity to thyroid hormones and its potential disruption of hormone activity by binding to thyroid hormone receptor sites. This merits further investigation given that triclosan has been found in breast milk, meaning that it finds its way into the body. Indeed, the chemical is so widespread in the environment that it turns up in the urine of the majority of the North American population. Of course detecting triclosan in the urine does not necessarily mean that we are at risk, although studies on mice and fish have shown a hindrance of heart muscle contraction at doses that are not far from human exposure.

Finally, there is the hypothesis that our overuse of cleaning agents and antimicrobials may be disrupting the human biome, that collection of some one hundred trillion bacteria that inhabit our body, outnumbering human cells ten to one. Some researchers believe that the increase being noted in the incidence of allergies, celiac disease, Crohn's disease, diabetes, mood disorders, obesity, and even autism is linked to a shift in the body's microbial environment.

The industry line is that triclosan is a "thoroughly researched chemical that has been safely used for decades." That is actually a hollow argument because the "thorough" research did not focus on the kinds of subtle effects that are raising eyebrows, and "safe use" is based on lack of acute effects. Indeed triclosan has no acute toxicity since its biological effect is based on the compound's ability to block a key bacterial enzyme that humans do not possess. While no specific health or environmental consequence has been linked to the widespread use of triclosan, it is unlikely that we would be worse off if it were removed from products where its claimed effectiveness to reduce bacterial disease has not been backed up by evidence. Our microbiome may even thank us.

SNAKES AND SNAKEROOT

Love can sometimes be deadly. Just ask Cleopatra. Of course you can't, unless you are someone like John Edward or James van Praagh, psychics who claim to be able to talk to the dead. The famed Egyptian queen committed suicide in 30 B.C. after hearing of the death of her great love, the Roman general Mark Antony. After the assassination of Julius Caesar, with whom Cleopatra had a relationship and a child, she teamed up with Antony to fight Octavian, Caesar's designated heir, in a battle to see who would rule Rome. Their joint armies were defeated by Octavian and when Antony heard a report that Cleopatra had been so devastated by the news that she committed suicide, he decided to follow suit. The report of the queen's demise turned out to be false.

When Cleopatra learned of Mark's death, she made up her mind to kill herself, but to do it without suffering. Supposedly she tested various poisons on condemned prisoners and concluded that the bite of the Egyptian cobra was lethal without inducing spasms of pain. It was also an appropriate way to go because the snake was an Egyptian symbol of divine royalty. The Egyptian cobra is often referred to as an asp, an anglicization of "apsis," an ancient term that referred to poisonous snakes found in the Nile region.

The queen had a servant smuggle the snake into the royal

chambers in a basket of figs and allowed herself to be bitten on the breast, or so goes the story. If that's the way it really went down, it was not in a painless fashion. The cobra bite induces tremendous swelling with excruciating pain as the skin stretches uncontrollably. The venom of the cobra has at least three different toxins; all three are proteins that either destroy cells or impair transmission of information from one nerve cell to another. Today antivenoms are available, produced by injecting a nonlethal dose of venom into a horse or sheep and then collecting the animal's blood to isolate the antibodies formed.

Just about the only animal that will tangle with a cobra successfully is the mongoose. This ferret-like animal supposedly prepares for its battle with cobras by dining on *Rauvolfia serpentina*, commonly known as Indian snakeroot. Legend has it that in India this plant was once used as an antidote for snakebite after observing that it was a big part of the mongoose's diet. Unfortunately, the legend has no basis in fact. Mongooses do suffer bites from snakes, however not often because the fangs of the cobra may not be able to reach the skin of the animal due to the thick fluffed fur. Furthermore, the mongoose is very quick and is highly skilled at attacking cobras and avoiding their strike. People are not that successful at avoiding bites. About 35,000 to 50,000 people die of snakebite every year in India.

Folkloric accounts often stimulate research to see if there is any validity to them, and sometimes there is. Ethanolic extracts of *Rauvolfia serpentina* do indeed neutralize cobra venom in the test tube, but there is no evidence that consuming the plant has any effect on snakebites. It can, however, have an effect on the mental state of a person. *Rauvolfia* is traditionally linked with the holy men and mystics of India, including the great spiritual leader Mahatma Gandhi. Such spiritual men have reportedly chewed the root of the plant to help them achieve a mental state of complete philosophic detachment during the meditation process. In Sanskrit, the plant is known as *chandrika*, which translates as "moonshine plant," because it was supposed to be a treatment for "moon disease," or mental illness thought to be related to the lunar phases. In this case there actually is something to the folklore.

Plants contain numerous compounds with potential biological activity, and snakeroot is no exception. Some fifty compounds with such potential have been identified, but the most interesting one is reserpine, first isolated in 1952. It made its mark in medicine mostly through the work of Dr. Nathan S. Kline (1916–1983), a graduate of the New York University School of Medicine. Kline is the only two-time winner of the Albert Lasker Award for Clinical Medical Research, an award sometimes referred to as America's Nobel Prize. In 1952, he headed a research unit at Rockland State Hospital in New York (later the Rockland Psychiatric Center) at a time when the national inpatient population in public hospitals was approaching the half-million mark. Traditional therapies seemed inadequate to treat the growing number of mentally ill patients.

Kline and his colleagues took the unusual step of investigating reserpine, which was already being used in the U.S. to treat high blood pressure. He was intrigued by stories from India about the calming effects of snakeroot. Trials with hospitalized patients found that approximately 70 percent of those suffering from schizophrenia were markedly relieved of their symptoms upon being treated with reserpine. This earned Kline his first Lasker Award.

Reserpine holds a place in history as "the original tranquilizer," and at one time was the only such medication employed as a calming drug on seriously disturbed psychiatric patients. But reserpine also comes with some undesirable side effects that can include edema and psychological problems such as nightmares and despondency, which can give rise to suicidal ideation. It has now been replaced by better drugs.

Encouraged by his success with this tranquilizer, Kline focused his research on drugs with a potential to treat mental illness. In 1964, he earned his second Lasker Award for the study and subsequent introduction of iproniazid, a monoamine oxidase inhibitor used in the treatment of severe depression. The successful use of drugs for two major categories of psychiatric illness led to the release of thousands who were able to rejoin society. Kline's work has been acknowledged as a major factor in opening a new era in psychiatry, namely psychopharmacology. And it all started with the folklore about the snakeroot plant, the mongoose, and the cobra.

SUGAR ISN'T SO SWEET

"Is it true that putting a piece of garlic in the rectum at night can cleanse the body?" And with that single question posed by an audience member back in 1975, my chemical focus shifted to food and nutrition. The question came after one of my first public talks on chemistry at a local library where I had described the role chemistry plays in our daily lives, mostly using dyes, drugs, plastics, and cosmetics as examples. I was sort of taken aback by the question but managed to stammer something like "Where did you hear that?" Back came the answer: "From *Panic in the Pantry*." After mentioning that my only experience with garlic had been with rubbing it on toast with some very satisfying results to the palate, I promised to check out the reference.

It wasn't hard to track down *Panic in the Pantry* in a local bookstore. The title had suggested some sort of attack on our food system but this turned out not to be the case. At least not in the way I had thought. Flipping through the book I came across terms such as "chemophobia," "carcinogen," "additives," "chemical-free," and "health foods." I was intrigued, especially on noting that the book had had been written by Dr. Frederick Stare, a physician with a degree in chemistry who had founded the Department of Nutrition at Harvard's School of Public Health, and his student Elizabeth Whelan. Within a day I had read *Panic in the Pantry* from cover to cover and was so captivated that I dove into the turbid waters of nutrition and food chemistry with great enthusiasm. Ever since then I have been trying to keep my head above water, buffeted by the growing waves of information and misinformation.

Panic in the Pantry focused on what the authors believed were unrealistic worries about our food supply, vigorously attacking the popular lay notion that "if you can't pronounce it, it must be harmful." Yes, that daft message was around long before the Food Babe made it her anthem. In truth, the risks and benefits of a chemical are a consequence of its molecular structure, and are determined by appropriate studies, not by the number of syllables in its name. Stare and Whelan also challenged the Delaney clause, a piece of U.S. legislation stating that no additive shall be deemed safe if it has been shown to cause cancer in any species upon any type of exposure. They pointed at studies that showed very different effects of chemicals in rodents and humans and maintained that it was unrealistic to condemn additives if exposure was not taken into account. "Too much sun can cause skin cancer, but does that mean we should stay indoors all the time?" they asked.

What about the curious case of the clove of garlic in the rectum? An excellent example of a misinterpretation of information, something that I have seen much too often. In a discussion of food faddism through the ages, the authors introduced the antics of one Adolphus Hohensee, who had forged a career as a "health food" advocate after his real estate business had landed him in jail for mail fraud. The dietary guru told his audiences that the sex act should last an hour, and if they did not measure up to this level of sexual adequacy it was because they had a diet laden with additives. He railed against these chemical scoundrels he claimed were promoted by the "American Murder Association" made up of the FDA, the Better Business Bureau, and food manufacturers. Not dissimilar to what some faddists assert today!

Hohensee's answer to the chemical onslaught was a clove of garlic in the rectum at night, with proof of its efficacy being the scent of garlic on the breath in the morning. Obviously the garlic had worked its way from bottom to top, cleansing everything in between. Far from promoting this regimen, Stare and Whelan had used it to highlight the extent of nutritional quackery.

I found most of the arguments in *Panic in the Pantry* highly palatable, but there was a discussion of one chemical that left a somewhat bitter taste. That chemical was sugar. I had been quite taken by *Pure, White, and Deadly*, a 1972 book by British physiologist Dr. John Yudkin that made a compelling case linking sugar to heart disease, cavities, diabetes, obesity, and possibly some cancers. Dr. Stare dismissed sugar as a culprit, implicating saturated fats as the cause of coronary disease. That to me seemed not to meet the standard of evidence that was applied to other issues in *Panic in the Pantry*.

As it turns out, there was a reason for Dr. Stare's dismissal of sugar as a health problem. In 1965 the Sugar Research Foundation (SRF), the industry's trade association, asked Stare to sit on its advisory board because of his expertise in the dietary causes of heart disease. The sugar industry was extremely worried about Yudkin's growing influence and had decided to embark on a major program to take the focus off of sugar and direct it towards fats. Stare's defense of sugar as a quick energy food that should be put in coffee or tea several times a day and calling Coca-Cola a healthy between-meals snack was welcomed by the industry.

As we have now learned from historical documents brought to light in a paper in the *Journal of the American Medical Association*, the SRF paid members of Stare's department to carry out a literature review, overseen by Stare, designed to point a finger at fats while expressing skepticism about sugar's supposed criminality. That review was published in the *New England Journal of Medicine* without any disclosure of sugar industry funding and successfully steered readers away from associating sugar with heart disease. While Stare was correct about many aspects of unfounded chemophobia, his reputation has now been tarnished by the undeclared payments received by his department from the sugar industry. Sugar, as we now know, is not as innocent as Dr. Stare had claimed. But at least he never did suggest garlic in the rectum to cleanse toxins. As far as I know, neither has the Food Babe.

BITTER ABOUT SUGAR

The play opened in 1960 at the Sullivan Street Playhouse in Manhattan to one of the most scathing reviews ever published in the *New Yorker*. "It is perfectly awful. It is supremely inept. It is magnificently foolish. It is sublimely, heroically, breathtakingly dreadful. It inspires a sacred terror. It is beyond criticism." Obviously critic Donald Malcolm didn't think much of *The Goose*, the brainchild of playwright Jerome Irving Cohen.

That name likely doesn't ring a bell, but perhaps that of J.I. Rodale does. After all, he was the founder of Rodale Inc., publishers of the popular *Prevention*, *Men's Health*, *Women's Health*, *Runner's World*, and *Organic Gardening* magazines. Cohen and Rodale are one and the same person. J.I. thought his birth name of Cohen was too Jewish, and when it came to business ventures, "Rodale" would resonate better with the public. The business ventures he had in mind were a push for organic agriculture and for the prevention of disease instead of puzzling over how to cure it.

He had some good ideas, such as cutting down on fat, sugar, and meat, eschewing tobacco, and emphasizing whole grains. His enthusiasm for dietary supplements was unsupported by evidence, but ads for such supplements certainly supported his publications. Rodale himself took some sixty supplements a day and on occasion even demonstrated their supposed efficacy in a somewhat foolhardy fashion. Once to demonstrate the efficacy of tablets he took to strengthen his bones, he deliberately threw himself down a flight of stairs. He wasn't unnerved by the sarcastic remarks this precipitated, maintaining he stood up to them fine because he took plenty of B vitamins that were good for the nerves. No wonder some called him a crank. Rodale, though, had a witty comeback. "Even the critics admit it takes a crank to turn things."

Turn the crank he did. Sometimes in the right direction, sometimes not. He deserves marks for promoting sustainable agriculture and for recommending eating fish and urging people to eat more berries and nuts. But he also maintained that kelp should be included in the diet because "the absence of hay fever cases in the Orient is due to the fact that the Japanese and Chinese eat liberally of this product." He also thought that wild game was a superfood because it was "free of the taint of chemical fertilizers." Fertilizers do not "taint" food. He might have had a point with pesticides, but not fertilizers.

The guru recommended that white sugar be dispensed with entirely, an idea that sounds sweet to many experts today. But he tainted his advice by saying that sugar should be replaced by maple syrup. This offers no significant advantage. Neither did his exhortations to include more honey in the diet make sense. Yes, it may have small amounts of antioxidants, but sugar is sugar, whether it comes from sugar beets, sugarcane, honey, or maple syrup.

It was Rodale's disdain for sugar that prompted him to pen *The Goose*, an attempt to promote his nutritional thoughts in a rather unconventional fashion. The play introduces John Gabriel, a juvenile delinquent in a Harlem housing project who runs with a gang and even steals money from his sister that he blows on soda pop, donuts, and the horses. His antics prompt a visit by a social worker who explains that juvenile delinquency is caused by an overindulgence in sugar. Her rationale? Consumption of sugar increased eleven-fold from 1900 to 1960 with a corresponding increase in crime rate. A classic confusion of association with cause and effect. Then comes a real "scientific" argument. "If sugar can cause cavities in the teeth," she asks, "what will it do to the brain, which is softer?"

Apparently the delinquent youngster is swayed by this powerful reasoning and decides to kick the sugar habit. He becomes a model youth, quits the gang, and returns his sister's money. Somehow this comes to the attention of "food and bottling interests" and one of their portly representatives shows up to offer John a bribe to get back on the sugar wagon. Apparently still morally weakened by residual sugar in his system, he accepts, and soon reverts to hoodlum behavior. Luckily the angelic social worker reappears and once more convinces him to forego ice cream, pop, and donuts, pulling him back from the brink of disaster. And with that the curtain falls. The audience files out with a memorable line still ringing in their ears: "Do you know mosquitoes never bite diabetics?" Apparently Rodale's message was that even mosquitoes are smart enough to stay away from sugar.

The harangue wasn't quite over. Brochures promising "The Proofs of Claims Made in the Play Called *The Goose*" were handed out in the lobby. The reader learned that Hitler could never get enough of his favorite whipped-cream cakes and that he was a sugar drunkard. "Was this what made him a restless, shouting, trigger-brained maniac?" The implied answer was "yes." Blaming Hitler's insanity on sugar was insane, but Rodale did bring up the issue of overconsumption of sugar long before it was fashionable to do so.

J.I. thought that following his dietary regimen and supplement intake would allow him to live to a hundred, unless, as he said, he was "run down by a sugar-crazed taxi driver." He didn't make it to a hundred, but it wasn't a taxi driver that did him down. It was a heart attack, one that happened in 1971 during a taping of *The Dick Cavett Show*, just after Rodale had bragged that thanks to his lifestyle he "had never felt better" in his life. The show never aired.

The Goose laid an egg, and not a golden one. It had a very short run, partly thanks to Malcolm's bitter review in which he promised to express his gratitude to the cast by assuring them that the secret of their identities was safe with him.

SUGAR CONSUMPTION

According to a New Guinean legend, the human race began when the first man made love to a stalk of sugarcane. Although conjuring up a mental picture of such an activity is intriguing, our love affair today is with the taste of sugar, not with the cane's physical attributes. It was indeed in New Guinea that sugarcane was first domesticated some ten thousand years ago, eventually spreading to Asia, where it became highly prized as a medicine to treat everything from headaches to impotence. By the Middle Ages, Arab tradesmen developed sugar refining into an industry that allowed the production of sweetened foods. Marzipan, made by mixing ground almonds with sugar, became popular and may well go down in history as the world's first "sugar added" food. Returning crusaders introduced sugar to European palates with demand soon outstripping supply. And that's when the sweet story of sugar took a bitter turn.

With the European climate not suitable for growing sugarcane, explorers sought more suitable locations. The Caribbean islands and Brazil were ideal, but the native population was not enough to supply workers for the mushrooming plantations. That set in motion one of history's darkest chapters, the African slave trade. Although mass slavery is now just an appalling memory of the past, there is still a connection to sugar. We have become slaves to its sweet taste, consuming an average of about twenty-five teaspoons of added sugar every day.

Until recently, the magnitude of that consumption lurked in the shadows as the spotlight focused on fats and cholesterol as dietary devils. But as it became clear that "low fat" was not the answer to obesity, diabetes, or cardiovascular disease, attention shifted to the ingredient that replaced fat in many processed foods, namely sugar. Pediatric endocrinologist Robert Lustig's YouTube presentation on the evils of sugar took the world by storm, and virtually every magazine and newspaper ran stories on the sugar hidden in our food supply and the "toxicity" of sugar-sweetened beverages. Two excellent films, the Canadian-made *Sugar Coated* and the Australian *That Sugar Film*, effectively laid out the health implications of overconsumption. The public also learned about the sugar and soft drink industries' attempts to prevent disclosure of the problems their own research had discovered. Shades of the activities of the tobacco industry.

My interest in sugar dates back to the early 1970s, when I came across a book by Dr. John Yudkin with the provocative title *Sweet and Dangerous*, featuring a sugar bowl anointed with the skull and crossbones on the cover. I must admit that I thought this was just another one of the many alarmist books that were appearing at the time, warning us to stay away from just about any food you could think of. We were being warned about fats, cholesterol, salt, additives, pesticide residues, and now sugar. But as I glanced at the qualification of Dr. Yudkin, I thought the book might be worth looking into. Yudkin was raised in London after his family had escaped the Russian pogroms in 1905 and went on to get both a PhD and an MD degree. His interest was nutrition, and he eventually became professor of Nutrition and Dietetics at the University of London. Clearly Yudkin was no scientific slouch.

In the 1950s, concern about the increasing rate of heart disease was addressed by Dr. Ancel Keys's famous Seven Countries Study in which he drew attention to a highly suggestive relationship between the intake of saturated fat, mostly from animal products, and the death rate from coronary disease. Yudkin was skeptical of Keys's analysis and claimed that sugar intake correlated with heart disease at least as well as fat intake. He also made some interesting observations, noting, for example, that the Masai in Kenya had a diet very high in animal fat but had very little heart disease. Many factors may be involved, but the Masai eat virtually no sugar. When Yudkin questioned patients who had a heart attack about their prior diet, he discovered that they had been eating about 120 grams of sugar a day, twice as much as his healthy controls. Unfortunately, Yudkin's work was relegated to the back burner by Keys's vilification of fats, enthusiastically supported by the sugar industry. That industry is now scrambling to deal with the assault on sugar as Yudkin's ideas are being vindicated and governments are urging people to limit their added sugar intake. Forty grams a day seems appropriate. Fewer sweets may make for a sweeter life.

SOME BEEFS WITH BEEF

These days a lot of people have a beef with beef. There is the ethical issue of raising animals for slaughter, an issue that cannot be settled by science. But science can certainly provide insight into the environmental impact of raising livestock, as well as into questions about the effects of eating beef on our health.

Raising cattle is not an environmentally friendly process, about that there can be no debate. First, there is the matter of cows burping methane, a greenhouse gas. Media reports will often mention that methane has a twenty-five times greater global warming potential than carbon dioxide. While that is true, it is misleading. Individually methane molecules do have twenty-five times greater warming potential than carbon dioxide molecules, but there are far fewer of them in the atmosphere. The concentration of carbon dioxide is about 400 parts per million (ppm) and methane is 1.85 ppm, so overall methane is certainly not responsible for greater global warming than carbon dioxide. However, the claim that livestock burps make up roughly 18 percent of all greenhouse gas emissions, more than what is emitted by all types of vehicles worldwide, appears to be correct. Then there is the amount of water needed to raise cattle. Every serving of beef requires about two thousand liters of water. Skip a burger and you can save enough water to shower with for at least a month.

While cattle usually start out grazing on grass or eating silage, they are "finished" on feedlots with corn or soy, grains that can also feed people. About 45 percent of all grain grown goes towards feeding animals. The oft-quoted statistic is that for cattle raised on feedlots, as is commonly the case in America, about 7 kilograms of grain are needed to produce 1 kilogram of meat. Obviously cows reared on grass and hay do not take food out of human mouths, but they do emit about twice as much methane as cattle raised on a feedlot, since they grow more slowly and live longer. The world has a lot of pastureland where crops can't be grown and calculations show that feeding livestock residues from other crops coupled with sustainable grazing could still provide about two-thirds of the meat that is currently produced. But given the world's growing appetite for meat, this is not likely to happen.

On the plus side, beef is an efficient way of providing protein relative to total calories. An adult requires roughly 50 grams of protein a day, an amount that can be provided by 200 grams of steak (400 calories). It would take 600 grams of kidney beans (760 calories) to provide the same amount of protein. Meat is also a good source of B vitamins, zinc, and iron. But there may be an issue with the iron found in meat.

Since the 1970s, researchers have noted a greater incidence of colorectal cancer in countries where beef consumption is high. The same does not hold for poultry or fish. What is the difference? One theory is that it is the iron content of beef, twice that of poultry, that may be responsible. Iron in meat is bound to heme, the non-protein part of hemoglobin, the compound that carries oxygen around the bloodstream. Heme iron is absorbed more readily than iron in plant products and laboratory studies show that this form of iron oxidizes fats to yield peroxyl radicals that can effectively cleave DNA, a possible initial step in triggering colon cancer.

Not all studies show an association between fresh meat consumption and colorectal cancer, but virtually all show a link with processed meats. That may have to do with the nitrites used as preservatives being converted into carcinogenic nitrosamines. In 2013 data from the European Prospective Investigation into Cancer and Nutrition study (EPIC) revealed that for every 50 grams of processed meat eaten per day, the risk of early death from all causes during the twelve-year period of the study increased by 18 percent. Researchers had followed half a million people, distinguishing between consumption of red meat, white meat, and processed meat, while controlling for factors such as smoking, weight, fitness, and education levels, all of which can influence health. When it came to fresh, unprocessed meat, there was no association with ill health. Indeed, people who ate no meat at all had a greater risk of early death than people who ate a little meat! Furthermore, the risk of early death was significantly reduced among meat eaters who ate a lot of fiber. In the U.S., the National Health and Nutrition Examination Survey (NHANES) has followed over 18,000 people for decades and has found no link at all between meat consumption and ill health.

The way meat is cooked can raise concerns. High temperatures produce heterocyclic aromatic amines, polycyclic aromatic hydrocarbons, and advanced glycation end products. Suffice it to say that these are nasty compounds, concentrated in charred parts of meat, that we can do without. Lower cooking temperatures are desirable, but then the problem of bacterial contamination rears its ugly head, particularly with burgers, since grinding can spread bacteria that normally are found only on the surface of meat throughout. Animals, like humans, harbor a host of bacteria, most of which are harmless. But not all. *E. coli* O157:H7, naturally found in the intestinal contents of some cattle, goats, and even sheep, can cause a condition known as hemolytic uremic syndrome (HUS) in which red blood cells are destroyed and kidney failure ensues. Back in 1993, four children died and 178 other people suffered permanent kidney or brain damage after consuming undercooked hamburgers at Jack in the Box restaurants in the U.S. This prompted an outcry for hamburgers to be cooked to an internal temperature of at least 160°F.

What do we distill out of all this? Eating a little red meat, less than 500 grams a week, in terms of health, is not likely to be detrimental. As long as it isn't charred! Vegans and vegetarians will point out, correctly, that it is possible to get all the nutrients one needs from a plant-based diet. However, there is also an enjoyment factor associated with eating, and to many people a hamburger is more tasty than a tofu burger. When it comes to the environment, however, the vegetarians win out. A kangaroo burger is a compromise since these animals do not produce methane. But let me tantalize you with a cricket burger. Seventy-five grams of these creatures will provide fifty grams of protein at a cost of only 340 calories. And they do not belch methane. However, there is that yuck factor to deal with.

PLASTIC PACKAGING PROS AND CONS

I went shopping for supper: a cucumber, cherry tomatoes, lettuce, bread, gnocchi, pasta sauce, chicken breast, blueberries, and Honeycrisp apples. The plan was to make chicken schnitzel with a side of gnocchi and a salad. Dessert would be fruit.

As I was unpacking my booty, something struck me. Aside

from the pasta sauce, which was in a bottle, everything else was packaged in plastic. The cucumber was wrapped in a polyethylene sleeve, my seven-grain bread probably in biaxially oriented polypropylene, the lettuce was in a resealable bag that I'm guessing was polypropylene. The cherry tomatoes and blueberries were in clear polyester containers, the gnocchi was vacuum packed in flexible plastic, likely composed of several layers of different polymers, and the chicken was sitting on a polystyrene tray, wrapped either in polyvinyl chloride or low density polyethylene. The reason for some of this guesswork is that while solid plastic containers must be identified using the familiar recycling triangle, plastic wrap requires no such labeling. My apples also made the journey in the common tear-off polyethylene bags provided in the supermarket.

As I surveyed the scene, three questions came to mind. Is all this plastic necessary? What is the environmental impact? Is there any issue with chemicals from the plastic leaching into food?

Packaging safeguards food from outside contaminants, protects products that are sensitive to oxygen exposure, reduces waste, and increases the variety of foods available. If not for proper packaging, blueberries or cherry tomatoes, replete with nutritional benefits, would not be available year-round. You don't see bins of blueberries and if you did, it would not be a pretty sight due to damage from handling and spoilage. That brings up the problem of food waste, a huge global issue. About a third of all food produced in the world, roughly enough to feed three billion people, is either lost or wasted. "Loss" occurs during production due to mold, insects, rodents, improper refrigeration, and inefficient transportation. "Waste" happens when food in restaurants gets discarded, when purchased food doesn't get eaten at home, or when produce in stores spoils or isn't bought because of some minor blemish.

Calculations show that in North America almost a pound of purchased food per person per day ends up being discarded as a result of spoilage and excessive table waste. Appropriate packaging can reduce waste both in the store and at home. For example, that plastic sleeve on the cucumber extends its shelf life from three to fourteen days by preventing moisture loss and contamination with airborne fungi. The biaxially oriented polypropylene ensures that no oxygen passes through to react with the polyunsaturated fats in my seven-grain bread. The gnocchi in the vacuum pack is virtually as fresh as the day it was packed because the multilayered plastic packaging, likely polyethylene and ethylene vinyl alcohol (EVOH), prevents moisture and oxygen from entering. Moisture is conducive to bacterial growth and oxidation of various food components results in off flavors. The tightly wrapped chicken was in no danger of contamination on its way from the store to my cutting board.

Clearly, plastic packaging can benefit retailers and consumers. But what about the environment? In an ideal world, all the plastic would be collected and recycled. But in the real world, 40 percent ends up in landfills, 32 percent ends up cluttering the landscape and the oceans. Of the 28 percent that gets collected, half is burned for energy and half is recycled. We have to do better than that. Estimates are that without significant action by 2050, there may be more plastic by weight in the ocean than fish.

Unfortunately, not all plastics can be recycled into their original form. Polyester can, which is why discarding polyester containers, including beverage bottles, is criminal since these can be processed into new containers. My cherry tomatoes came in a container made from recycled polyester. Multilayer plastics cannot be readily recycled because their components cannot be separated, but they can still be recycled into lower value items such as plastic lumber.

Chemists are hard at work developing single recyclable polymers that have the flexibility, barrier properties, and strength of multilayer plastics. There is also extensive research aimed at high performance polymers that can reduce the amount of plastic needed for packaging. For example, polyethylene can be coated with a layer of silicon dioxide, essentially glass, that is only one millionth of a meter thick, giving it barrier properties similar to the multilayer plastics. The reduced weight means a much smaller carbon footprint. While paper, aluminum, and glass are more readily recycled than plastics, much less plastic is needed to do the same job, meaning that transportation costs, both financially and environmentally, are reduced.

Of course the environmental issues have to be weighed against reducing food waste because waste also has an environmental impact. Just think of the cost of seed, the agrochemicals, and the energy required to process and transport food that never gets eaten.

Obviously, one can debate the environmental pros and cons of plastic packaging ad nauseam. But what about effects on our health? Plastics can harbor unreacted monomers as well as a variety of catalysts, antioxidants, and plasticizers used in manufacture. One could paint a dire picture, for example, of estrogenic phthalate plasticizers migrating into food from PVC cling wraps. Without a doubt traces of these chemicals can be detected, but the amounts are trivial. Ditto for other components such as antimony trioxide used as a catalyst in polyester manufacture.

In my view, the benefits of plastics used in food packaging outweigh the detriments, mainly due to enhanced food safety, greater variety, and less waste. But we do have to emphasize recycling and reduction where possible. I brought my purchases home in a reusable bag. And I did some good for the environment by reducing the need to manufacture money. I paid with plastic. And if you are interested, the schnitzel and gnocchi were yummy, the salad fresh, the blueberries tasty, and the apple crisp. That's because it had been protected with a wax spray. But that's another story for another time.

PLASTIC PROBLEMS

Science can make for a strange bedfellow. I had just finished recording a video showing off one of my favorite sweaters and expounding on the ingenuity and the environmental benefit of it being made from recycled polyester bottles when an article appeared on one of my newsfeeds about how "your clothes are poisoning our oceans and food supply." The message was that the very fabric I was so high on may be unraveling the fabric of society.

I must say I was puzzled by the headline, but on glancing through the story, the details of the problem quickly came out in the wash, as it were. Synthetic fabrics are not exactly inert; they release microscopic bits of fiber when washed. The particles may be microscopic, but their number is anything but. Researchers at the University of California found that a synthetic fleece jacket releases hundreds of thousands of microscopic fibers, about 2 grams in total, with each wash. Wastewater treatment removes some of this debris, but most of it ends up in rivers, lakes, and oceans where it can be consumed by wildlife. The fibers then can bioaccumulate up the aquatic food chain, right up to people consuming fish. Whether this presents a risk is not known, but bits of plastic are not a desirable dietary component. The clothing industry is sensitive to the problem and is working on coatings for fabric that would reduce shedding. Also in the works are washing machines that prevent the release of microfibers by using pressurized carbon dioxide instead of water.

The shedding of microfibers from synthetic fabrics is not the only way tiny pieces of plastic, invisible to the naked eye, end up in water systems. Microbeads, introduced into consumer products such as toothpaste and exfoliating skin products as abrasives, are a bigger concern. Six varieties of the tiny beads are used. Those composed of either polyethylene, polypropylene, or expanded polystyrene are more likely to float, whereas the ones made of polyvinyl chloride, nylon, or polyethylene terephthalate (PET) are more likely to sink. McGill biologist Anthony Ricciardi has found microbeads in significant numbers in sediment at the bottom of the St. Lawrence River, meaning possible contamination of fish that feed on the riverbed.

Microbeads range in size from 10 millionths of a meter to one millimeter. Their round shape makes them much less irritating than irregularly shaped, abrasive exfoliants such as apricot kernels or walnut shells that have sharp edges. Also, because the particles are tiny spheres, they act as little ball bearings, allowing for easy spreadability of creams and lotions as well as a smooth texture and silky feel. There's more. Imperfections in the skin tend to be visible because of the contrast between how they reflect light compared with the surrounding tissue. Microbeads with their ability to scatter and diffuse light can minimize the appearance of fine lines and even out skin tone. When it comes to toothpaste, though, they make a minimal contribution to polishing the teeth and may actually become embedded in gum tissue. Why are they there? Since the microbeads can be produced in various colors, they can also increase the visual appeal of a product.

A single container of face wash can contain hundreds of thousands of the microspheres. While the virtually indestructible plastic beads are not themselves toxic, once they enter the water, they attract potentially toxic substances such as PCBs, triclosan, and nonylphenols. Like the microfibers, microbeads can then become part of the aquatic food chain, eaten by fish and then by people. Once consumed, the beads may also leach out plastic additives such as colorants, plasticizers, and ultraviolet light stabilizers.

Researchers have found fish contaminated with microbeads in both the oceans and the Great Lakes. Besides carrying toxins, the beads can cause internal abrasions and can stunt growth of fish by giving them a false sense of being full. One-third of fish caught off the southwest coast of England have been found to contain microbeads, and Belgian researchers studying seafood from German farms and French supermarkets found that an average portion of mussels can contain about ninety microplastic particles and an order of oysters about fifty. The beads have also been found to lodge in the guts of crabs as well as in their gills.

The number of microbeads that end up in the environment is staggering. In New York State alone some nineteen tons go down the drain every year. Most wastewater plants are not equipped to filter out such fine particles and while they could be retrofitted, the expense would be prohibitive. Drinking water poses less of a problem because municipal water treatment plants can filter out the tiny particles, although a sampling of German beers found microbeads in every bottle, with the water used being the likely source. Both Canada and the U.S. have moved to ban microbeads and manufacturers have started the process of phasing them out. Researchers agree that there are still too many unknowns to fully assess the environmental damage caused by microplastics, but given that they do not contribute significant benefits, they should be eliminated.

But the problem of plastic waste in the oceans is greater than can be accounted for by microfibers and microbeads. Other tiny particles form from the breakdown of plastic bags, bottles, and all sorts of containers that get discarded and end up in waste streams that empty into the ocean. "Biodegradable" on a label means that the plastic has been shown to degrade under ideal composting conditions, but these do not exist in the natural environment. Estimates are that the ratio of plastics to fish by weight in the oceans is 1:5, and with our current callous attitude towards "reduce, reuse, recycle," it is set to increase to 1:1 by 2050.

Given these concerns, I don't think I can wear my "made from a plastic bottle" sweater with the same pride as before. And I may even feel a bit of apprehension tossing it into the laundry basket.

BLANKETS, BALLOONS, AND SPACE SUITS

"Throw that baseball as hard as you can!" That's what we would tell an audience volunteer back in the 1980s when with colleagues David Harpp and Ariel Fenster we gave presentations on plastics at Man and His World, an exhibition that was an offspring of Montreal's famous Expo 67. The target was a Mylar emergency blanket that we would unfold and hold tight several feet in front of the "pitcher." But before the volunteer would unleash the missile, he and the rest of the audience would be treated to a discussion of polyester film and its metalized versions.

The concept of linking certain small molecules together to produce long chains of polyester was developed in the 1930s by DuPont chemist Wallace Carothers. But polyester research took a back seat when Carothers came up with nylon, a more commercially viable polymer. However, British chemists John Whinfield and James Dickson pursued Carothers's work and in 1941 developed a polyester fabric eventually marketed as Terylene. Then in 1946, DuPont bought the legal rights to polyester manufacture and began to market polyester as the magical fabric that needed no ironing. This was soon followed by Mylar, a polyester film that could be made thirteen times thinner than a human hair and still absorb the impact of a baseball thrown at up to eighty miles per hour. Since it was unlikely that any of our volunteers could muster such a speed, the only thing we had to worry about was being hit by the ball. Luckily that never happened.

Polyester films can also be coated with a thin layer of metal, usually aluminum. This reduces the permeability of the film and gives it a reflective surface, just the properties needed for construction of the world's first "satelloon." In 1960, Project Echo placed a 30.5-meter-diameter metalized Mylar balloon into Earth orbit to reflect telephone, radio, and television signals between continents. Mylar films were also used in NASA's spacesuits for radiation resistance and for keeping astronauts warm. Later the technology was applied to produce emergency blankets to conserve body heat for shock victims, premature babies, and marathon runners.

Mylar found application in food packaging and decorative items such as Christmas tinsel. And the helium balloon industry took to aluminized polyester because it kept the gas from diffusing out as readily as from a rubber balloon. Mylar balloons became popular at celebrations of all sorts without people giving much thought to where they eventually end up. But it is time to give the issue some thought.

A good number of these balloons escape and can cause two types of mischief. They can float into power lines and because of the conductivity of the aluminum can cause outages and even explosions. The problem is not trivial: San Diego Gas & Electric recorded 312 outages in the last five years caused by balloons. As a consequence, a California bill now proposes to ban Mylar balloons. Florists, decorators, and party supply stores are up in arms, claiming that millions of dollars in business would be lost. California actually has a current law requiring Mylar balloons to be weighted down so they can't fly away. The balloon industry claims that a ban is overkill and the problem of rogue balloons could be eliminated if the law were properly enforced.

Some municipalities in England are also considering a ban, but there the concern is that whatever goes up must eventually come down. And the balloons can come down in the ocean, where the metallized polyester breaks down into small pieces that can be mistaken for food by fish.

Runaway Mylar balloons are a legitimate problem, but other allegations made about polyester don't hold water. Like the claim that drinking from a polyester water bottle that has been left in a hot car is a risk factor for breast cancer. That warning has been circulating ever since singer Sheryl Crow appeared on *The Ellen DeGeneres Show* to discuss her breast cancer diagnosis. When such a calamity occurs, people commonly look for possible causes of their affliction and the discussion turned to water bottles left in a hot car. The lay press is full of articles demonizing plastics, mostly in an unjustified fashion, and Sheryl likely heard about chemicals like bisphenol A and the phthalates that have been accused of causing cancer. Neither of these is present in the kind of water bottles Sheryl was concerned about. Bisphenol A is used to make polycarbonate plastics, which are used in the large carboys that sit on top of water coolers but not in the commonly used water bottles, which are made of polyester. And here some confusion about phthalates enters the picture.

One of the chemicals used to make polyester is terephthalic acid, trace amounts of which could conceivably leach into the water. The hot car connection comes from the correct notion that an increase in temperature increases the leaching of water-soluble compounds from plastic. But terephtalic acid is not the phthalate that has been implicated by some in health problems. That is diethylhexylphthalate, a chemical added to some plastics, mostly to polyvinyl chloride (PVC), to increase their flexibility. The only PVC used in water bottles is in the cap, any leaching from that would be trivial. And it needs to be emphasized that there is no credible evidence that bisphenol A or phthalates play a role in breast cancer.

The real problem with polyester bottles is that too many are discarded instead of being recycled. Bottles can be processed into pellets that when melted can be used to make items ranging from plastic lumber for benches to fibers for clothing and carpets. And Japanese researchers have even floated a new balloon. They have found a bacterium, *Ideonella sakaiensis* 201-F6, that can biodegrade polyester into its original building blocks that can then be used to make new items without relying on petroleum.

BAGGING PLASTIC BAGS?

Demystifying science is a mystifying experience. You quickly learn that every issue is more complicated than it first seems. Whether you are exploring bisphenol A, fluoridation, phthalates, GMOs, dietary supplements, nonstick cookware, preservatives, tap water, pesticides, or climate change, you find a variety of opinions ostensibly supported by references plucked from the scientific literature. But with the enormous number of papers being published it is easy to pick and choose references that support almost any agenda. The quality of scientific publications follows a bell curve: some are outstanding, some are dismal, most are mediocre. Any painting of controversial issues, and there is controversy about virtually every issue, as either white or black should be regarded with suspicion. Science comes in various shades of gray, with the hues shifting as new information comes to light. Risks, and there always are some, have to be evaluated in relation to benefits.

Of course not all disputes are of equal importance, and often even the importance of disputes is disputed. Disposable plastic bags are a case in point. Some look on them as a gigantic environmental scourge that has to be eliminated while others believe that plastic bags constitute a minor environmental woe and efforts to eliminate them amount to using a jackhammer to attack an ant. There's no question that plastic bags are a symbol of our throwaway culture and are an inviting target for scorn because they are a visible sign of pollution. They can be seen fluttering from trees, floating in that much-publicized patch of plastic detritus in the middle of the Pacific Ocean, and clogging sewers in parts of Asia. But the bags don't dive into the ocean, jump into sewers, or take flight without help. Human help. We are the real problem. With proper recycling, reuse, or disposal, benefits can outweigh risks.

What then are the perceived risks? Arguments usually revolve around the bags being made from oil, a nonrenewable resource, the plastic being nonbiodegradable, the bags taking up space in landfills, the bags being unnecessary because of ready replacement by paper or reusable bags, and the bags leaving a large carbon footprint. Disposable bags are made of high-density polyethylene (HDPE), which is manufactured from ethylene derived either from petroleum or natural gas. In Canada, the source is usually ethylene made from ethane, a component of natural gas that otherwise is commonly burned off.

Plastic bags do not biodegrade in a landfill, as we are often told. This is true, but modern landfills are designed to have a low oxygen environment to prevent biodegradation that would result in the formation of methane, a greenhouse gas. The purpose of a landfill is to seal in the contents and prevent substances from leaching out. Since plastic bags are highly compressible, they take up very little volume in landfills. In any case, plastic shopping bags are estimated to make up less than 1 percent of litter.

Paper shopping bags do not biodegrade in a landfill either, and because of their greater mass they are a greater burden on the waste stream. Paper manufacture is an energy-intensive process and requires the use of many chemicals. Cradle-to-grave calculations generally show that plastic bags have a lower carbon footprint than paper bags. "Biodegradable" bags are a marketing scheme; they don't degrade under normal conditions.

But why should we make an issue of plastic versus paper? Why not rely on reusable bags? Here too, the issue is not as simple as it seems. A cotton bag would have to be used about 130 times in order to have a carbon footprint that is less than that of a plastic bag. Growing cotton requires more pesticides than most crops, and processing and transport require a great deal of energy. If the plastic bag is reused to line your garbage can, a cotton bag would have to be used over 300 times to have a lower global warming potential.

Reusable plastic bags are often made of laminated plastics and are not recyclable. Depending on the type of plastic, whether low-density polyethylene (LDPE) or nonwoven polypropylene, a reusable bag would have to be used at least ten to twenty times before it becomes more environmentally friendly than a disposable bag. There is also the issue of contamination if reusable bags are not cleaned properly. A warm trunk is an excellent incubator for bacteria originating from that trace of meat juice left in the bag.

If not reused for that next trip to the grocery store, or for lining garbage bins, or for collecting garbage in a car, or for picking up after pets, or for covering food in the fridge, disposable plastic bags are eminently recyclable into plastic lumber, trash cans, containers, and new plastic bags.

Many municipalities and even countries have banned the giveaway of plastic bags or have introduced fees for them. This has resulted in the use of more paper bags, not an environmental plus, and an increase in the sales of plastic bags for garbage bins.

The publicity given to the plastic bag problem is out of proportion to its size. When it comes to energy expenditure, producing a hamburger requires about forty times as much as a plastic bag. And that hamburger is not always totally eaten. Food waste makes up ten times as much of the waste stream as plastic bags. When it comes to environmental concerns, what you put inside the shopping bag is more important than the type of bag.

While charging for plastic bags in order to reduce waste and encourage reusable bags is sensible, a total ban causes unnecessary inconvenience. What are you going to do if you've forgotten your reusable at home and have to stop at the store for your evening salad? Tomatoes in one pocket and peppers in the other?

BPA RESEARCH – WHEN IS ENOUGH ENOUGH?

"More research is needed." That's a common final sentence in scientific papers, especially when it comes to studying the effects of environmental chemicals on health. With numerous chemical reactions going on in our body all the time, and exposure to thousands and thousands of chemicals, both natural and synthetic, it is a huge challenge to tease out the effects of a single substance. That brings up the question of when the effort and funds invested in studying one chemical is enough. Is there a point at which further research is unlikely to lead to a major revelation? Can research funds be better spent on alternative projects that are more likely to yield meaningful results? We may be reaching such a stage with bisphenol A (BPA), a chemical that has been the subject of more studies in the toxicological literature than any other.

BPA was quietly cruising under the radar as a component of polycarbonate plastics, dental sealants, thermal receipt paper, and epoxy resins until 1995, when a paper by Dr. David Feldman of Stanford University entitled "Estrogens in Unexpected Places: Possible Implications for Researchers and Consumers" put it squarely in the spotlight. Feldman had been investigating whether a certain yeast solution was capable of producing estrogens and found that it did indeed activate estrogen receptors in rat uterus tissue. But further studies showed that the substance responsible for the activity did not come from the yeast, rather it leached from the polycarbonate flasks used during the autoclaving procedure. This stimulated a great deal of interest because of the possibility of BPA being ingested after leaching from food containers and epoxy resin linings in cans.

Frederick vom Saal of the University of Missouri was one of the researchers who became interested in BPA. He had been studying mice treated with estrogen during pregnancy and found an increase in the prostate size of their male offspring. Now he wondered if BPA, with its estrogenic properties, would also have such an effect. It did, and vom Saal raised the alarm because he claimed the doses used were similar to what humans might encounter. Although others were unable to reproduce vom Saal's results, his work unleashed a plethora of studies. Within the last twenty-five years, more than 8,000 studies dealing with various aspects of BPA have been published. One would think that such a massive effort would have resulted in some solid conclusions about the health effects of BPA, but such is not the case.

Numerous researchers have confirmed BPA's estrogenic activity, but found it to be orders of magnitude less than estradiol, the estrogen produced naturally in the body. There is also good agreement that orally ingested BPA is rapidly eliminated in the urine. Other than that, one can cherry pick studies to show that BPA is associated with obesity, developmental problems, infertility, cancer, heart disease, and diabetes. Or pick ones that show no link with any of these conditions. Regulatory agencies such as Health Canada, the Food and Drug Administration (FDA) in the U.S., the European Food Safety Authority, and many others have carefully scrutinized the data and have concluded that at common exposure levels BPA does not pose a risk. But some scientists who have built careers on demonizing BPA disagree and urge that all possible measures be taken to reduce exposure. They also clamor for more studies to be carried out, although it seems unlikely that the 8,001st study will somehow clarify the murky situation presented by the 8,000 previous ones.

Nevertheless, the studies keep coming. Researchers like to jump on a hot topic, especially one that gets a lot of media attention. But just how much the emerging data adds to the body of knowledge that already exists is questionable. One recent study concluded that female mice exposed to BPA were more lazy at night than controls. It is hard to see its relevance to us. In another case, researchers calculated how much BPA students ingest in school cafeterias. Consuming pizza, milk, fresh fruit, and vegetables was associated with minimal amounts of BPA, while the same meal with canned fruits and vegetables delivered up to 1.19 micrograms per kilogram of body weight. But even this is less than half the dose deemed safe by the most cautious regulatory agencies. Incidentally, BPA is never present in polyester, the plastic used for bottled beverages.

Given that the major regulatory agencies do not believe the population is exposed to unsafe levels of BPA, and that significant efforts are already underway to find substitutes for the chemical, it seems appropriate to funnel research efforts and funds towards more pressing needs.

PLASTINATION CONTROVERSY

Dr. Gunther von Hagens never appears in public without a black fedora. He wears it in tribute to Dr. Nicolaes Tulp, who was Amsterdam's official city anatomist in the seventeenth century. Tulp is depicted, black hat and all, in one of Rembrandt's most famous paintings, *The Anatomy Lesson of Dr. Nicolaes Tulp*, in which the anatomist is shown pointing out some of the features of a cadaver to a group of fascinated onlookers. Such public dissections were popular events back in the seventeenth century, attended by students as well as the general public. The corpses usually were those of executed criminals with the one in Rembrandt's painting being that of Aris Kindt, convicted of armed robbery and hanged the same day the anatomy lesson shown in the painting took place.

So who is Dr. Gunther von Hagens? He's the German anatomist who invented the technique of plastination for preserving biological specimens, including whole human bodies. The

"Body Worlds" exhibits he developed have toured the globe, educating people about the workings of the human body by allowing spectacular views of the insides of plasticized cadavers. The educational aspects of the exhibit are clear-cut, but the concept is mired in controversy. Some religious groups oppose the public display of dead bodies in this fashion, and questions have arisen about the origin of some of the corpses, although von Hagens maintains that the bodies are always obtained legally.

Whatever ethical issues may arise about plastination, there is no question that it is a spectacular scientific achievement. Attempts to preserve human bodies date all the way back to the ancient Egyptians and mummification, but never before has it been possible to keep the anatomy intact to the degree made possible by plastination. Von Hagens deserves a great deal of credit for working out the necessary chemistry.

First, the body is treated with formaldehyde to kill all bacteria and prevent decomposition. Immersion in freezing acetone then draws water out of the tissues and replaces it with acetone. A bath in a mix of a silicone polymer and dibutyltindilaurate catalyst is next. As vacuum is applied, the acetone that had permeated the tissues boils off and the silicone is drawn into the vacated space. Under the influence of the catalyst, the silicone chains join together end-to-end, increasing the viscosity of the solution.

The next step is exposure to the vapors of tetraethoxysilane, a catalyst that forms bridges between adjacent silicone chains, resulting in a sturdy three-dimensional network. The bodies are then ready to be posed and displayed, mainly thanks to silicone chemistry. But preserved bodies are far from the only place we find silicones. You will find them in hair products, face creams, sealants, adhesives, waterproof coatings, insulators, cooking utensils, lubricants, defoaming agents, dry cleaning solvents, baby bottle nipples, sex toys, bracelets, Silly Putty, and a host of medical devices. Obviously, the silicone industry is extensive, and the need for raw material for silicone synthesis is huge. Luckily there is no shortage, since the starting material for the production of all silicones is silica, SiO_2, found in nature as quartz or sand.

The first medical application of silicones occurred in the 1940s and was based upon the water-repellent property of these compounds. When blood contacts glass or metal it tends to form clots that can clog needles, syringes, and transfusion equipment. Researchers discovered that a coating of silicone prevented blood from clotting, and today it is common for blood-collecting apparatus to be coated with these chemicals. In 1946, silicone found another use when American surgeon Frank Lahey repaired a bile duct with a piece of flexible silicone tubing. Silicone turned out to be biocompatible and biodurable and soon became the prime material for catheters and was even used to replace diseased male urethras.

Perhaps the most interesting of the early uses of silicone was in the production of a hydrocephalus shunt. Hydrocephalus is a condition whereby cerebrospinal fluid builds up in the brain and can cause problems if not drained. In 1955, a baby was born with a rare neural tube defect that required the implantation of a polyethylene shunt catheter to drain excess fluid from the brain into the heart. This required a valve to open to allow the fluid to drain when pressure built up and close to prevent back flow when the pressure equalized. A scaled-down version of an automotive pressure release valve was designed but it often became clogged. As it turned out, John Holter, the baby's father, was mechanically inclined and constructed a valve from some flexible tubing and two rubber condoms that seemed to fit the bill. However, the valve didn't stand up well to autoclaving, which was needed for sterilization. On approaching a rubber company for advice, he was told about silicone being a thermally stable material. The Dow Corning Company provided some free samples of silicone for experimentation and by 1956 the desperate father had produced a workable valve. Unfortunately by then his son was too ill for the implantation, but his surgeon, Dr. Eugene Spitz, implanted the valve in another child suffering from hydrocephalus. The newfangled ventriculoatrial shunt was so successful that its mass production began almost immediately and the Holter valve is still being used today. Perhaps it will appear in one of Dr. von Hagens's plastinated bodies.

Von Hagens sees the human body as a beautiful, exquisite machine and believes his exhibits are an excellent way to teach people about anatomy. He is fervently anti-smoking and claims that his display of a blackened smoker's lung has turned many visitors off smoking. Eventually von Hagens will put his body where his mouth is. He suffers from Parkinson's disease and will have his body plastinated so that people can learn about the condition. The unanimated anatomist will then welcome visitors to the Body Worlds exhibit, of course sporting his black fedora.

THE RISE OF BAKING POWDER

The label on the baking powder said "aluminum-free" and "GMO-free." What gives? Why would baking powder contain aluminum or GMOs, and why would we want to eliminate these substances? Let's start with what baking powder is, and what it does.

Yeast has been used since antiquity to make dough rise. But it takes time for yeast to generate carbon dioxide, and that isn't ideal for cakes, biscuits, or muffins. Baking powder by contrast goes to work as soon as it is wetted. The chemistry is quite simple. Carbonates or bicarbonates release carbon dioxide gas on reaction with an acid. Historically, the first substance used for leavening without yeast was potassium carbonate known as pearl ash, extracted from the ashes left behind when wood burns. Combining this with an acid such as sour milk or lemon juice caused the release of carbon dioxide gas. This bit of chemistry was introduced to cooks in 1790, but there was a problem. Soaps form when fats react with alkalis, and potassium carbonate is an alkaline substance. Cakes and biscuits developed a soapy taste! The solution was to use sodium bicarbonate, which has much less of a tendency to react with fats. But its use still required the addition of an acid. A breakthrough came in 1843, and it was all because Elizabeth Bird was allergic to yeast.

Luckily Mrs. Bird was married to a chemist. A clever one. It seems that Mrs. Bird was plagued with allergies and reacted to eggs as well as to yeast. But she loved custard. That had driven her husband into the lab to try to formulate an eggless version of the treat. He eventually found that a finely powdered mix of cornstarch, salt, and vanilla flavoring produced an acceptable custard mimic when mixed with water. Yellow natural coloring from the annatto plant completed the illusion. Bird's eggless custard is still popular in the U.K. and is traditionally served with "spotted dick." That's a "pudding" made with raisins, which are the spots. The "dick" part apparently stems from "puddick," a corruption of the word pudding.

After solving Mrs. Bird's custard cravings, her husband turned his attention to the yeast problem. His idea was to combine sodium bicarbonate with a solid acid to produce a mixture that was inactive until the components dissolved in water. Once in solution, the bicarbonate and the acid would react to quickly generate carbon dioxide. Tartaric acid, available from wine sediment, fit the bill. But there was another problem. When stored, the mixture would slowly absorb moisture from the air and release carbon dioxide prematurely. Then Bird had another brainstorm. Why not add some substance that would absorb any excess moisture? He hit upon starch. Now Mrs. Bird could have her baked goods without having to worry about a reaction to yeast. But she had to get her batter into the oven very quickly before all the carbon dioxide from the dough escaped.

That problem would plague baking powder manufacturers for years. And there were plenty of manufacturers, each one trying to get a step up on the competition. In the U.S., chemist Eben Horsford suggested replacing tartaric acid, increasingly hard to acquire, with another solid acid, monocalcium phosphate, that could be produced from bones. This led to the establishment of the Rumford Chemical Company. Its measured proportions of monocalcium phosphate, sodium bicarbonate, and starch became Horsford's Bread Preparation, eventually named Rumford Baking Powder. These days the monocalcium phosphate used in baking powders is produced from calcium phosphate that is mined.

Rivalry between manufacturers was intense and Royal Baking Powder rose to the top by sinking huge amounts of money into advertising. But fortunes changed when others discovered that aluminum sulfate was a cheaper acid than Royal's tartaric. Royal, however, could not switch because the company had signed a long-term contract with tartaric acid manufacturers.

It was aluminum sulfate that eventually solved the problem of having to work quickly with a batter. In 1889, the Calumet Baking Powder Company introduced double acting baking powder that featured an acid that reacts with the baking soda quickly, and another that reacts only when the batter is heated in the oven. In this way the release of the carbon dioxide is prolonged and a more fluffy texture can be achieved. Sodium aluminum sulfate and sodium aluminum phosphate are compounds that become acidic when heated and were incorporated into double acting baking powders.

It is, however, possible to have a double acting baking powder that has only one acid component. Monocalcium phosphate, when used in the right proportion with baking soda, reacts to release carbon dioxide and in the process is converted to calcium monohydrogen phosphate, which becomes acidic when heated. This is the double acting baking powder that can be labeled as "aluminum-free."

Why is that regarded as something positive? Some people claim that they can detect a metallic taste if aluminum compounds are present, and some worry because of a purported connection between aluminum and Alzheimer's disease. There is not much evidence for aluminum being a significant factor in Alzheimer's, but in any case, aluminum is the most prevalent metal in the earth's crust and is found in trace quantities in almost every food. The amount contributed by baking powder is not significant.

What about the "GMO-free" claim? That has to do with the starch that is a part of baking powder. Besides absorbing moisture, it makes pouring easier and provides more volume for easier measurement. The starch can be derived from potatoes or corn, and if it is from corn, it is possible that the corn is genetically modified. This is a nonissue since the cornstarch is highly purified and does not contain any genetic material. Declaring baking powder to be GMO-free may soothe the worrywarts but it has no practical significance.

IKARIAN LONGEVITY

The Greek island of Ikaria is named after Icarus, who according to mythology tried to soar like a bird with wings he had fabricated out of feathers. Unfortunately, he plunged into the sea after the wax with which he had attached the wings melted as he flew too close to the sun. While Icarus had a short life, modern Ikarians boast of impressive longevity, and unlike the inhabitants of the Hunza Valley of Pakistan who claim extreme longevity, Ikarians have birth certificates to prove their age. On Ikaria, ten times as many people live to the age of ninety as in the rest of Europe. They have lower rates of heart disease and cancer, suffer less from depression and dementia, and apparently maintain an active sex life into old age. The question is, what are they doing right? According to researchers at the University of Athens, a factor in their longevity may be the Greek coffee they drink.

Scientists used ultrasound to study how arteries in the arm respond to changes in blood flow as they applied and then released pressure through a cuff on the arm. How effectively the arteries dilate after being constricted is a measure of the health of the artery and a determinant of cardiovascular risk. Because of contradictory studies on the links between coffee consumption and heart disease over the years, cardiologist Dr. Gerasimos Siasos and colleagues thought that assessing the coffee consumption and arterial health of senior Ikarians could prove to be fruitful. In a sample of seventy-one male and seventy-one female subjects over the age of sixty-five, he found that chronic Greek coffee consumption was associated with improved endothelial function.

The endothelium is the inner lining of blood vessels and produces the chemicals that dilate or constrict the vessels. In endothelial dysfunction there is an imbalance in these substances leading to cardiovascular disease. Subjects who consumed mainly Greek coffee had better endothelial function than those who drank other types of coffee. Most drank 200 to 450 milliliters a day, with the beneficial effect being proportional to the amount consumed. Greek coffee made by boiling finely ground coffee differs in composition from other coffees. Coffee beans contain hundreds of compounds and the boiling process is more efficient at extracting these than other methods of preparation. Based on laboratory experiments, some of these compounds, cafestol for example, may increase blood cholesterol; others such as kahweol and an array of polyphenols may protect against cancer and heart disease. It is therefore theoretically possible that different methods of preparing coffee may have different health effects.

As one might expect, the study generated headlines around the world like "Greek coffee may help you live longer," and "Greek coffee may be the key to long life." That may well be so, as long as you live on the island of Ikaria and follow the lifestyle of the Ikarians. There is a lot more to their impressive longevity than drinking Greek coffee. What that might be is of course of great interest. Diet is a prime consideration. Ikarians eat little meat, little refined sugar, and lots of olive oil. They eat six times more beans than Americans and lots of locally grown greens that are especially high in antioxidants. You won't find many processed foods or soft drinks in Ikarian households. So the answer to their longevity may lie not only in what they are eating or drinking, but also in what they are not consuming.

Then there is the Ikarian habit of taking afternoon naps, the benefit of which is bolstered by a study of Greek adults that showed a 40 percent reduction in the risk of heart disease associated with regular napping. Most Ikarians admit to drinking a couple of glasses of red wine a day, which may be an underestimate. They drink goat's milk as well as local "mountain tea" made from herbs such as sage, marjoram, mint, rosemary, and dandelion. Some of these have diuretic properties that may lead to a slightly lower blood pressure throughout Ikarians' lives. Honey is often taken as medicine.

There's more. Ikarians walk a lot on their hilly terrain, dig and plant in their gardens, and take part in many festivals that involve dancing through the night. They are active people. They don't go home at night to sit on the couch and watch TV. They are very social and have close relationships with friends and neighbors. As one of the few physicians on the island points out, Ikaria is not a "me" place, it's an "us" place. Almost nobody wears a watch. And nobody takes dietary supplements.

What appears to link all the places in the world where unusual longevity has been documented is a mostly plant-based diet and social structure. It is easy to take naps, eat vegetables, and dance at festivals when everyone else does the same thing.

How long the people of Ikaria will maintain their longevity will likely depend on the extent of the encroachment of a western lifestyle. On Okinawa, the Japanese island also noted for longevity, this is already happening. Okinawans have gone from a diet of a low 1,800 calories a day, centered on sweet potatoes, vegetables, beans, and small amounts of fish, to one based on fast foods with a corresponding decline in longevity. Seventh Day Adventists on the other hand maintain their traditional totally vegetarian lifestyle and have seen no such decline.

The bottom line here? Remember that the coffee study only measured epithelial function. The assumption is that improved function leads to greater longevity. But the researchers did not determine if the subject with better epithelial function actually lived longer. I think we can safely assume that substituting Greek coffee for your morning Starbucks is not going to increase your life expectancy. Substituting an Ikarian type of lifestyle for our western one, however, might.

BOOSTING BRAINPOWER

We live in a world where people attempt to settle differences with insane wars and acts of terror. We foul our nest with an array of pollutants. We succumb to preventable diseases. We struggle to feed a growing global population. We clearly need to be smarter! If only there were a pill to boost our brainpower! A pill to allow us to learn things more quickly and actually remember what we learned. A pill to allow us to focus on whatever task is at hand without permitting our mind to drift. A pill for a better functioning brain. A "smart drug." A "nootropic."

Nootropics are chemicals that improve mental function in some way. The term was coined in 1972 from the Greek words for "mind" and "turn," by Corneliu Giurgea, a Romanian-born pharmaceutical researcher who was working for the Belgian drug manufacturer UCB. Giurgea was trained both in psychology and chemistry and had an interest in drugs that would improve cognition, that is, those that would "turn the mind" in the right direction. In the 1960s, while tackling the problem of motion sickness, he investigated gamma-aminobutanoic acid (GABA), one of the many neurotransmitters nerve cells use to communicate with each other. GABA cannot be taken by mouth because it does not cross the blood-brain barrier, so Giurgea focused on synthesizing analogs that could be administered orally.

He came up with piracetam, a compound that did not help with motion sickness but did reduce nerve cell excitability, a property that proved to be useful in the treatment of myoclonus, the sudden, involuntary twitching of muscles, commonly afflicting the arms or legs. But Giurgea also found that the drug improved cognition, leading to piracetam being commonly described as the "first synthetic nootropic." It hit the market in Europe as Nootropil, a name obviously chosen to imply a nootropic effect. The stated indication was for the treatment of myoclonus, but once a drug is approved for sale, physicians can prescribe it as they see fit, and piracetam soon developed an off-label use as a brain function enhancer. It also stimulated research into analogs, particularly in the old Soviet Union, with phenylpiracetam hitting the market as an anxiety-reducing and memory-enhancing agent. The Russians also developed phenibut, another GABA analog that was used by cosmonauts to reduce stress and improve focus during space missions.

Describing piracetam as the first synthetic nootropic is not exactly correct. That distinction should go to amphetamine, also first synthesized by a Romanian chemist. Lazar Edeleanu made the compound in 1887, although its effects on the mind were not recognized at the time. By World War II, though, both the Allies and the Germans were using amphetamine for its performance-enhancing effects, with pilots commonly popping the pills to stay alert during missions. Today, amphetamine is the active ingredient in Adderall, a medication used to treat attention deficit hyperactivity disorder (ADHD). It is also used and abused by students studying for exams because of its ability to increase focus.

Long before the advent of synthetic chemistry, many natural substances had developed a reputation as "brain enhancers." The most widely used such substance is caffeine, which can certainly wake up a brain. But it doesn't make you smarter. Olive leaf extracts have long been used, prompted by the biblical passage found in Ezekiel 47:12: "The fruit thereof shall be for meat, and the leaf thereof for medicine." Research on rats indicates that oleuropein, a compound found in olive leaves, may actually have an effect on cognitive function. When researchers treated rats with oleuropein, the usual cognitive dysfunction induced by administering colchicine directly into the brain did not occur. How do you know if a rat is cognitively dysfunctional? It has more difficulty in learning to find its way through a maze. So if your rats are having memory problems, you might think of treating them with a dose of olive leaf extract.

The herb *Bacopa monnieri* was used by Ayurvedic physicians in India 1,500 years ago to improve intellect, and gotu kola was recommended for memory enhancement. Traditional Chinese Medicine (TCM) has a number of brain boosting herbs, with ginkgo biloba leading the way. In South America, maca root and yerba mate tea have a long-standing reputation as focus enhancers. Cocaine was once promoted by Sigmund Freud as a cognitive enhancer. Not a wise choice. All of these, save for cocaine, are available today as supplements and are often hyped as natural nootropics, but the evidence for any significant effect is scanty.

Today, the most talked about nootropic is modafinil, trade name Provigil. The drug was developed in France in the 1980s as a treatment for narcolepsy and is now used for sleep and shift work disorders, but it is often used off-label as a cognitive enhancer. U.S. Air Force physicians prescribe it to pilots for fatigue management and the drug is available to the crew of the International Space Station to optimize performance while fatigued. Like any drug, modafinil can have side effects that range from nervousness and insomnia to anxiety and a potentially very serious rash. Modafinil received a boost for its much-hyped nootropic effect from the movie *Limitless*, since the fictional drug NZT in the film was thought to be modeled on modafinil. In *Limitless* the drug endows the hero with extreme mental prowess, far beyond anything achievable by nootropics.

When Dr. Corneliu Giurgea was asked back in the 1960s why he was so interested in drugs that would improve mental function, he replied that "man is not going to wait passively for millions of years before evolution offers him a better brain." So far, in spite of a great deal of research focused on prospective brain-enhancing drugs, we are still waiting for that better brain as evidenced by goings on in the world.

CADMIUM DANGERS

Shrek the friendly ogre delighted audiences in the 2010 movie hit *Shrek Forever After*. But for fast food giant McDonald's, Shrek turned out to be a nightmare. As a cross promotional feature, the company introduced a set of glasses decorated with images of Shrek and other characters from the film. After millions of the glasses had been sold, a problem cropped up that led to a large-scale recall. The yellow pigment used on the cups turned out to be cadmium sulfide, a substance toxic even in small amounts. The concern was that the pigment might rub off on children's hands and end up being ingested if they then put their hands into their mouth.

Cadmium was discovered in 1817 by Professor Friedrich Strohmeyer in Germany while looking into a problem encountered by apothecaries who were making calamine lotion for skin care. The process involved heating "calamine," a natural ore of zinc carbonate, to produce zinc oxide, which is the active ingredient in calamine lotion. Sometimes the lotion would end up with a yellow discoloration which Strohmeyer determined was due to a mineral contaminant that he eventually identified as a compound of cadmium.

It was the color of cadmium compounds that led to their first commercial use. Artists loved the bright yellow of cadmium sulfide and the reds and oranges resulting from a mixture of cadmium sulfide and cadmium selenide. Vincent van Gogh used cadmium sulfide to impart the yellow color to his flowers in his famous *Flowers in a Blue Vase*. Unfortunately, with time, cadmium sulfide oxidizes to cadmium sulfate, which is white, resulting in the original color of the painting being slowly altered. Claude Monet's famous yellow hues were also achieved with cadmium pigments.

Cadmium paints are still used today, although they are being phased out. Indeed, Sweden has submitted a report to the European Chemicals Agency claiming that artists rinsing their brushes in the sink are responsible for spreading cadmium over agricultural land via sewage sludge.

Cadmium is a cumulative toxin and the World Health Organization has suggested 70 micrograms as the maximum daily safe intake. Ingesting some cadmium is unavoidable because it shows up in crops. How does it get there? Sewage sludge and phosphate rock, both used as fertilizer, can harbor cadmium. As a result, a hamburger can contain about 30 micrograms of cadmium that can be traced to the grass or hay the cow ate, and ultimately to the soil in which the feed was grown. Coal also contains cadmium compounds that can end up in the atmosphere, from where they find their way into soil via rain. Other cadmium compounds may also be released from the nickel-cadmium battery industry, although modern pollution control methods minimize such losses. Cadmium can be also be found in significant amounts as a contaminant in zinc ores and some is released into the environment when the ore is mined as well as when it is smelted into zinc.

Nobody actually carried out a study to determine how much cadmium pigment can rub off onto little hands when gripping a Shrek glass, but it could well be less than what is found in the hamburger those hands are clutching. Still, eliminating any avoidable source of cadmium is desirable, especially since there is suspicion that cadmium compounds may be carcinogenic. Cadmium can also build up in joints and the spine causing a disease that the Japanese have named Itai-Itai, which translates as "ouch-ouch," due to the painful sounds made by victims as cadmium accumulates.

A classic case of environmental cadmium toxicity can be traced back to the early 1900s, although its cause was not identified until the 1960s. It was obvious that something was going on in the vicinity of the Jinzu River in Japan and its tributaries in China. People were getting sick, screaming in pain, and dying prematurely. Suspicion fell on the river and the mining companies that for years and years had been disgorging their wastes into the water. The mountains upstream were rich in minerals that contained silver, lead, copper, and zinc, and mines had been operating there for centuries. As demand for these metals increased in the twentieth century, more and more mining wastes found their way into the river, including increased amounts of cadmium ores.

River water was used for irrigation of rice fields, and since rice absorbs cadmium effectively, the metal accumulated in the food supply and consequently in the bodies of the population. The result was ouch-ouch disease. Although cadmium was only identified as the cause around 1965, by the late 1940s it had become obvious that the disease was linked to the water supply, and mining companies began to store their wastes instead of releasing them into the river. This prevented more people from contracting cadmium poisoning, but nobody really knows how many victims the mining operations created since they began to pollute the Jinzu River back in the sixteenth century.

In 1966 in England, a construction worker died and several others were sickened as a result of inhaling cadmium fumes. The men were using a welding torch to remove bolts as they were dismantling a construction tower used in the building of a bridge. It is common practice to electroplate steel bolts with cadmium, particularly those exposed to water. This is especially useful when there is contact with sea water since cadmium reacts with salt to form an impervious layer of cadmium chloride. In this case, the men inhaled the cadmium vaporized by the heat of the welding torch and suffered an acute reaction.

Shrek glasses are not the only items aimed at children that have caused a concern about cadmium. With lead being non grata, cadmium has been turning up in jewelry aimed at young girls, mostly originating in China. If pieces are accidentally swallowed, or if the jewelry comes into frequent contact with the mouth, enough cadmium may enter the bloodstream to cause harm. Jewelry made with cadmium should go the way of the Shrek glasses.

A MATTER OF TASTE

When it comes to food, everyone has likes and dislikes. Chocolate generally gets favorable comments, spinach less so. But no flavor seems to elicit the degree of polarizing comments as that of cilantro. There are websites and Facebook groups dedicated to demonizing cilantro, likening its aroma to soap or, curiously, to dead bugs.

The seeds of the cilantro plant are known as coriander and are even mentioned in the book of Exodus. Archeologists found some in King Tutankhamen's tomb, perhaps placed there with hopes of adding some spice to the afterlife. The ancient Chinese believed there would be no need to worry about the afterlife if you consumed cilantro because the herb conferred immortality. Hippocrates used it as medicine and even today some people ascribe health benefits to the herb based on its content of antioxidants, antibacterial compounds, and minerals. These, though, are not unique to cilantro. All plants contain varying quantities of these substances.

Another supposed benefit is cilantro's ability to chelate heavy metals. The term "chelate" comes from the Greek meaning "claw" and refers to compounds that have the ability to remove harmful metal ions from solution by gripping them like a claw. Some bloggers even push cilantro as an ingredient in a "detox" salad, claiming it rids the body of heavy metals. As usual, there is a kernel of truth to the claim, but that kernel is inflated with nonsense until it pops.

A few studies have shown that cilantro leaves can produce a chelating effect in water spiked with heavy metals and that cilantro can reduce absorption of lead when food tainted with it is fed to mice. But these effects are light-years from a salad with cilantro accomplishing any sort of heavy metal "detoxing" in people. Such a claim would require a demonstration of there being a heavy metal problem in the first place and its reduction with cilantro. A PubMed search for "cilantro detox" yields zero entries. Similarly, there is no basis to some food faddists' claim that "cilantro can reduce waterweight, is a cancer fighter, and can improve memory with its brain-protecting vitamins and minerals."

While the scientific literature provides no evidence for health benefits, it does provide clues when it comes to cilantro's polarizing flavor. What we refer to as flavor is the sensation triggered when molecules in food encounter receptors on our taste buds and in our nasal passage. Indeed, scent is an integral part of the sensation as evidenced by cilantro haters not being bothered if they consume the herb while holding their nose.

Some forty compounds have been isolated from cilantro, including a number in the aldehyde family that are mainly responsible for the aroma and taste. The composition of the seeds is somewhat different, having linalool, also found in lavender and cannabis, as a major component. It has a pleasant floral scent, accounting for its use in cleaning agents, detergents, and shampoos. When inhaled, it can reduce stress. At least in lab rats. Rats that inhaled linalool saw a reduction in the elevated levels of white blood cells induced by stress.

It is the aldehydes in cilantro that cause some people to liken the scent to soaps and lotions because these compounds are indeed found in those products. But why only some people? One theory is that the cilantrophobes are "supertasters" and can taste compounds that others can't. Supertasters do exist, but they react to very specific bitter compounds such as propylthiouracil, while most people taste nothing. However, there are no such compounds in cilantro, and supertasters are no more likely to be cilantro haters than anyone else.

It seems, though, that people who abhor cilantro may have some sort of genetic connection, if we go by an interesting study carried out by Dr. Charles Wysocki of the Monell Chemical Senses Center in Philadelphia. Taking advantage of the annual twins festival in Twinsburg, Ohio, Wysocki had identical and fraternal twins rate the scent of chopped cilantro. There were definitely lovers and haters, with identical twins almost always agreeing with their sibling, which was not the case for fraternal twins. Experiments at Monell have also separated the components of cilantro using gas chromatography and showed that while everyone can smell the "soapy" aldehydes, cilantro haters cannot smell the compounds that make the herb so attractive to its fans.

Interestingly, there is also an ethnocultural connection. A study at the University of Toronto surveyed 1,639 young adults and had them rate their preference for cilantro on a nine-point scale. East Asians were the most likely to dislike cilantro, with roughly 21 percent expressing their distaste. Caucasians were not far behind at 17 percent. Only 14 percent of those of African descent disliked the taste, followed by South Asians at 7 percent, Hispanics at 4 percent, and Middle Eastern subjects at 3 percent. These stats roughly parallel the use of cilantro in the cuisine of these areas, suggesting that there is a connection between liking cilantro and frequency of exposure.

While cilantro's enemies would rather stick rusty needles into their eyeballs than eat the fresh herb, they normally don't object to cilantro in prepared foods such as pesto. That's because the herb's flavor changes as the volatile aldehydes escape into the air when the herb is crushed, cooked, or pureed. Cilantro fans of course crave fresh cilantro and when cooking add the herb at the end stage. As for me, I'm with Julia Child on this one. Back in 2002, she told Larry King in an interview that if she found cilantro in a dish she was served she would pick it out and throw it on the floor. I recognize, though, that there are people who would jump to catch it before it hit the ground because they just love the smell and taste of this herb that has pleased some and irritated others since biblical times.

SPREADING KINDNESS

How does a rotten potato in a German concentration camp lead to a highly successful snack bar business in the U.S.? For one, it takes an unusual act of kindness by a Nazi guard.

Dachau, the first concentration camp opened by the Germans, housed many Lithuanian Jews, including Roman Lubetzky. The treatment of the prisoners was brutal, and hunger in the camp was rampant. But it seems not all the Nazi guards were monsters. One of them, on seeing Lubetzky's plight, threw him a rotten potato that provided some sustenance and, more importantly, offered hope. Lubetzky survived the camp, and after the war, he ended up in Mexico, eventually moving to the U.S. His son, Daniel, was tremendously moved by his father's experience and was inspired to do whatever he could to prevent such atrocities from happening again.

What better place to start a peace movement than building bridges between Jews and Arabs in the Middle East? So Daniel started a business selling a sun-dried tomato spread made collaboratively by Jews and Arabs. In the process he introduced the idea of a not-only-for-profit business that would eventually spawn a snack bar empire under the KIND trademark. The name was a tribute to his father, whose life was a model of compassion towards others and whose survival in the concentration camp was linked to an act of kindness. The idea for a snack bar came while Lubetzky was working long days after founding PeaceWorks, an organization aimed at forging cooperative relationships between warring peoples. He traveled a lot and had trouble finding a healthy snack on the road. Why not try to create one that was wholesome, traveled well, tasted good, and could also conform to a mission of spreading kindness? Thus the line of KIND bars was born with the idea of "being kind to your body, kind to your taste buds, and kind to your world."

KIND bars are basically made of nuts, fruit, seeds, grains, and honey, ingredients that according to the company "you can see and pronounce." Judging by the growth in sales since the bars were introduced in 2004, they certainly satisfy the taste buds. As far as "kind to your world" goes, Lubetzky's KIND Movement donates $10,000 a month to community projects. But the "being kind to your body" concept proved to be niggling since America's Food and Drug Administration (FDA) didn't take kindly to the claim of "healthy" on the label. It seems the FDA has requirements that need to be satisfied for use of the term. The food must contain less than 5 grams of total fat, less than 2 grams of saturated fat, and less than 480 milligrams of sodium per serving. It must also contain at least 10 percent of the Daily Value for vitamins A, C, calcium, iron, protein, and fiber.

The problem was that some of the KIND bars contained more saturated fat than allowed under the "healthy" claim. There was also an issue with the plus sign used on the label to designate bars with extra antioxidants, fiber, or protein. A definition applies here as well. To bear a plus sign, the bar has to contain 10 percent more of the nutrients than a bar the FDA has deemed representative of the snack bar category. Some KIND bars did not meet this requirement.

Were the labels really goading people into eating an excess of saturated fat under the guise of a "healthy" claim? No. Actually recent studies have failed to demonstrate a link between saturated fats in the diet and heart disease. Even more significantly, the saturated fats in this case come from the nuts in the bar, and nuts have actually been shown to lower the rate of heart disease.

What we have here is a case of overzealous nut-picking by the FDA, but not nearly as overzealous as the class action lawsuit against the manufacturer launched in California prompted by the FDA action.

The suit alleges that consumers were misled by the "healthy" claim as well as by the "non-GMO" and "natural" designations on some labels. The plaintiffs maintain that some ingredients, such as soy lecithin, are derived from GM soy, and furthermore, that they are highly processed, including extraction with solvents such as hexane. Consumers who were deceived should be compensated, the suit claims, and the manufacturer should be made to pay punitive damages. Sure sounds to me like a frivolous lawsuit and abuse of the legal system.

As if that weren't enough of a nuisance, the U.S. military has issued an advisory that its personnel should not consume certain KIND bars. This time it has nothing to do with supposed false claims. The bars in question contain hemp seeds! Hemp is in the same botanical family as marijuana, but it has been bred to contain very little tetrahydrocannabinol (THC), the psychoactive ingredient in marijuana. Hemp seeds are nutritious, containing all nine essential amino acids and mostly heart-healthy polyunsaturated fats with an omega-6 to omega-3 ratio of 2 to 1, a ratio considered to be ideal for health. The seeds are also rich in vitamin E, phosphorus, potassium, magnesium, iron, and zinc, all important for proper immune function.

So what is the problem? The seeds contain trace amounts of THC, far too little to have any effect, but possibly enough to show up on some drug screening tests. Army regulations ban the use of any hemp product so a soldier testing positive would be in trouble, hence the warning about KIND bars that contain hemp seeds. Of course, soldiers can safely consume any of the other KIND snacks. For the rest of us, hemp seeds are fine. They're even kind to Canada. Growing hemp in the U.S. is illegal, but this is not the case in Canada. Curiously, Americans are allowed to import hemp seed, which KIND does.

OATS VS. POP-TARTS

So what's for breakfast? A bowl of steel-cut oats with mix-ins. Wild blueberries, sour cherries, bananas, slivers of almond, ground flax or chia seeds, and cottage cheese. I could tell you that it is because of the cholesterol-lowering beta glucans in the oats, the antioxidant properties of the anthocyanins in blueberries, the anti-inflammatory cyanidin in the sour cherries, the potassium in bananas, the magnesium and fiber in almonds, the omega-3 fats in the seeds, and the protein in the cottage cheese. But the truth is that I eat this not because I have succumbed to orthorexia nervosa, the pathological fixation on eating proper food, but because I really like the taste. And of course nutritionwise it sure beats Pop-Tarts.

Yes, some people do breakfast on that ultrasweet pastry with no redeeming nutritional value. It apparently does have financial value though, enriching the Kellogg's coffers by some $800 million a year! While companies, including Kellogg's, are stumbling over each other to concoct "natural," "additive free," "gluten free," "fat free," and "sugar free" cereals, the sales of Pop-Tarts have not been impeded by their 30 percent sugar content or their long list of additives. That 30 percent sugar translates to about 18 grams, or almost five teaspoons. The World Health Organization recommends that our intake of sugar be limited to six teaspoons a day, so after eating just one Pop-Tart there isn't much wiggle room. Why do people eat these things? Maybe it's nostalgia, a longing for those carefree days back in the 1960s when Pop-Tarts were first introduced, or maybe it's just the convenience of grabbing a Pop-Tart and coffee before dashing out the door in the morning. There isn't even a need for a toaster. Pop-Tarts can be eaten raw. Convenient, for sure. Healthy? Not exactly.

Of course eating in a proper fashion is important, but that does not mean every morsel ingested has to be evaluated in terms of being healthy or unhealthy. But that is the message that is being purveyed by popular scientifically unsophisticated and sanctimonious nutritional gurus who believe that if you don't regularly dine on non-GMO, "chemical-free" food, you are doomed. The problem isn't that railing against processed food has no merit, the problem is the implication that if you stray from fresh raw organic vegetable juice, goji berry cookies, or kale massaged with kimchi, or if you dare to cook your quinoa in anything but an "organic clay pot," you are heading straight to health purgatory. Of course you "can discover what is truly healthy" by subscribing to a monthly food guide from the "Babe" who has used her nonexistent nutritional background to carry out extensive "investigations." The result of this naïve guruism is an irrational fear of food that has been deemed to be unhealthy and an unhealthy preoccupation with food that has been blessed by this high priestess of salubrious eating.

Given the minimal processing involved, I suspect my breakfast of steel-cut oats would get a blessing. Basically the whole grain is lightly toasted to develop a nutty flavor and inactivate enzymes that could lead to rancidity. Then the grains are cut into smaller bits by rotating steel blades in order to increase their surface area. That means the oats will absorb water more easily and cook more quickly than intact grains, but they still require twenty to thirty minutes to cook. They also require frequent stirring to prevent the formation of clumps, traditionally with a spurtle, a wooden stick designed for this purpose.

The time and the need to stir put some people off, so the industry has come up with rolled oats. In this case, the grains are steamed to make them soft and pliable, and then are flattened by passing them between rollers. They will absorb more liquid and cook faster than steel-cut oats, but they also tend to become mushy when compared with the texture of the steel-cut variety. This is the version usually added to granola bars, cookies, and muffins.

Even the slight cooking needed here is too much for some folks who do not want to do more than pour hot water onto their cereal. Enter instant oats that are precooked, dried, and rolled really thin. They absorb water quickly and are really ready instantly. The price of convenience is a mushy texture and a more bland taste. As far as nutritional content goes, the difference between steel-cut, rolled, and instant oats is insignificant. But there is a difference when it comes to their effect on blood sugar. Steel-cut oats are digested more slowly, they make you feel full longer, and they have a lower glycemic index, meaning they have a smaller impact on blood sugar.

There is one more difference that can arise with instant oats. Some have lots of added salt and sugar. McDonald's Fruit & Maple Oatmeal has 32 grams of sugar, more than that in a bag of M&M's! Some instant oats also have calcium carbonate or guar gum added as thickeners. There's nothing really wrong with these, but they are indicative of a more highly processed product. I'll stick to my minimally processed steel-cut oats, with no additives except for what I choose to include. I've become pretty good at wielding a spurtle, and I may even think about entering the Golden Spurtle competition held each year in Scotland for the most outstanding porridge. I understand the current champion is Bob's Red Mill Organic Steel-Cut Oats. They claim to be ready in five to seven minutes but do not reveal how this is accomplished. Perhaps by presteaming. I'll have to do a little scientific investigation. It's always a good experiment when you can eat the results. And a final confession. I have on rare occasions eaten a Pop-Tart. Without any feelings of guilt.

EMULSIFIERS ON TRIAL

The headline caught my eye: "Food Preservatives Linked to Obesity and Gut Disease." It wasn't on the front page of some tabloid or on some scientifically illiterate activist's blog. It was featured on the website of *Nature*, one of the world's premier scientific journals, atop an article summarizing the findings of a study published in the journal. Intrigued, I looked up the study, finding that it was entitled "Dietary Emulsifiers Impact the Mouse Gut Microbiota Promoting Colitis and Metabolic Syndrome" and had nothing at all to do with preservatives. One would think that copywriters working for *Nature* would be familiar with the difference between preservatives and emulsifiers. But I digress. Let's deal with the study.

If your food came in a box, a jar, a bottle, a can, or some sort of plastic container, it's been "processed." That means it was somehow milled, flaked, puffed, sweetened, salted, colored, flavored, pickled, steamed, baked, fortified, pasteurized, chemically preserved, dehydrated, or emulsified. And chances are it has also been criticized, its cacophony of additives and chemicals leached from packaging blamed for causing a host of ailments that include cancer, heart disease, diabetes, hyperactivity, developmental problems, and obesity. Until the *Nature* paper, emulsifiers had been cruising under the alarmists' radar.

Food production often involves the need to attain a proper consistency by combining components that normally do not mix with each other. An oil and water mixture, such as peanut butter, is a classic example. Enter an emulsifier! Without it, an oily layer separates and floats on top, making for an unappetizing product. Emulsifiers are molecules that feature an oil-soluble end as well as a water-soluble one and essentially form a link between fat and water to prevent their separation. There are numerous emulsifiers, ranging from monoglycerides derived from fats that can keep oil and vinegar from separating in a salad dressing, to carrageenan, extracted from seaweed, that keeps the insoluble cocoa particles suspended in chocolate milk. Emulsifiers are also common ingredients in commercial mayonnaise, ice cream, bread, and pastries.

Basically, the *Nature* study reported that mice fed a diet containing either of two common emulsifiers, carboxymethylcellulose or polysorbate 80, at concentrations that can be found in our food supply, had blood sugar control problems and an increased risk of gaining weight and of developing inflammatory bowel disease. The explanation offered is that emulsifiers can disrupt the mucus membrane that lines the intestine, allowing bacteria to penetrate and cause inflammation as they contact the intestinal wall. The weight gain may be caused by changing the bacterial composition of the gut in favor of species that break down food readily and allow greater absorption of its components. An interesting piece of research to be sure! The authors suggest that "the broad use of emulsifying agents might be contributing to an increased societal incidence of obesity and metabolic syndrome and other chronic inflammatory diseases." Of course the key word is "may."

While the concentration of emulsifier in the animals' diet was comparable to that found in, for example, a salad dressing, it is not the concentration that matters. It is the total amount ingested that counts. Our diet does not consist solely of salad dressing, so the total amount of emulsifier the mice consumed on a continuous basis was far more than human exposure. Also, there are numerous emulsifiers used in our food supply; others may not have the effects seen here. There is also the issue of natural emulsifiers such as lecithin in eggs. Might these have a similar effect? As is almost always the case, we have to conclude that "more research is needed." Certainly the evidence that disruptions of our gut flora can affect health in various ways is mounting, and given that emulsifiers are most commonly found in processed foods that are shrouded in other nutritional issues, we have yet another reason to emphasize cooking with fresh ingredients instead of looking for convenience.

Then, just as we are set to cast a wary eye on foods with emulsifiers, researchers at the University of Nottingham come out with the suggestion that emulsifiers may be the answer to the problem of obesity! Using the right emulsifier, they claim in a paper in the *British Journal of Nutrition*, can double the time food takes to leave the stomach, and consequently can stave off hunger pangs longer than usual. The idea is that when water separates from food, it leaves the stomach much more quickly, but when it is prevented from escaping by an emulsifier, the food stays bulkier and people feel more full, less hungry, and have less appetite. But it does have to be the right emulsifier, one that doesn't break down on exposure to stomach acid. Sorbitan monostearate is the one the researchers studied.

Volunteers were asked to drink a milkshake type of concoction made with olive oil and water, flavored with coffee to make it palatable. The fullness of their stomach was monitored with a scanner, and they were also questioned about their sense of fullness, appetite, and hunger at hourly intervals for twelve hours.

After one hour, there was twice as much volume in the stomachs of the subjects who had consumed the mixture with the stable emulsifier. The stable emulsion meal also made subjects feel fuller and less hungry than those who had consumed the chemically similar but less stable polyethoxylated sorbitan monostearate.

What do we make of all this? Not a whole lot. The first study with mice on a diet was not representative of what humans consume and the second, while it demonstrated a temporarily curbing of appetite, did not actually show weight loss. But I wouldn't be surprised to see hucksters start promoting sorbitan monostearate as the new weight loss miracle.

SHAKE SHAKE

I'm going to offer some unsolicited advice to Beachbody, the multilevel marketing company that made its name by selling exercise videos for home use. But first a look at "21 Day Fix Extreme," the company's popular program that aims to turn flab into fit in just twenty-one days with intense exercise, dietary advice, and a clever system of plastic containers for portion control. And then there's Shakeology, an assortment of flavored powders used to make shakes that are an integral part of Fix Extreme.

Shakeology claims to help you lose weight, reduce junk food cravings, provide healthy energy, and support digestion and regularity. Those are kind of vague, but whenever I come across the nonsensical claim, that it "helps alkalize the body," like Elvis, I get all shook up. Ditto for the meaningless "promotes detoxification" and "boosts the immune system" claims. What Shakeology's shakes shake down to is a clever exercise in marketing.

Protein has an aura of health, since it is common knowledge that it is the basic component of muscle. So when trying to create a "superfood formula," another hollow term, it makes sense to start with some protein, never mind that the average person gets more than enough in their daily diet. Although whey protein is perfectly fine, it sounds too ordinary, so better to include some esoteric sources like chia and quinoa. If you can, find a protein source that nobody has ever heard of, like sacha inchi, even better. To be fair, there is some evidence that a single large dose of protein may curb hunger, but the source of that protein is irrelevant.

After protein, proceed to throw everything other than powdered kitchen sink into the mix. Vitamins? Sounds healthy but a deficiency is very unlikely for Shakeology customers. If they can shell out $120 for some protein powder, they likely have a pretty varied diet. Next, scour the scientific literature for any herbs, plants, berries, antioxidants, microbes, or enzymes that have ever been shown to have any potential benefit in some laboratory or animal study and toss them into the formula so they can appear in the list of ingredients. Ignore the fact that the amounts added are way, way less than those used in the studies.

Blueberries, spinach, acai berries, and pomegranate are the darlings of the antioxidant worshippers, so add some sort of extract of these to garner attention, and hope that nobody asks "how much?" Ginkgo biloba has been associated with memory improvement, albeit controversially, so include an insignificant amount, and while you're at it, why not add some fungi such as *Cordyceps* with an undeserved reputation as an aphrodisiac. There has been a lot of positive discussion about probiotics, so be sure to include some "beneficial" bacteria, but stay silent about the inadequate numbers. And of course don't forget to mention the inclusion of kale, the current dietary sweetheart. Hope that nobody asks about how much is included.

Shakeology's creator Carl Daikeler explains that his product is completely different and works in a "more sophisticated way, supporting eleven systems in the body, and giving the body the power to heal itself." Meaningless humbug. The real question is whether Fix Extreme delivers on its promise.

The company has apparently carried out one clinical trial with fifty people, lasting ninety days, that never made it into the peer-reviewed literature. Details are murky, but apparently some meals were replaced with shakes, which is not the way the program is currently marketed. The noted weight loss and cholesterol reduction was attributed to Shakeology. That's an unwarranted assumption since the effect could well have been due to reduced calorie intake. And since there was no control group using another protein source, it cannot be assumed that the Shakeology shakes are effective. But we can assume that if a dedicated person cuts down on portion size, as will be the case by using Beachbody's containers, and exercises like the dickens, they will lose some weight over the twenty-one days of the Fix Extreme regimen. I must say I have been impressed by the success of my daughter and son-in-law. They have religiously abided by the program, lost weight, and claim the shake tastes so good that there are no visions of chocolate cake dancing in their heads.

Now for the advice to Beachbody I promised. Cut the flimflam about the shake's superfood ingredients, alkalizing, and immune-boosting properties and concentrate on promoting portion control, exercise, and how the shake is a tasty replacement for sweet snacks. Finally, I think KC and the Sunshine Band's 1976 hit "(Shake, Shake, Shake) Shake Your Booty" would be a nice fit for Fix Extreme ads.

BONING UP ON COLLAGEN

"Everything old is new again," so goes the 1970s song. And had Professor Justus von Liebig not left us in 1873, he would be singing along, reveling in all the praise being heaped upon bone broth these days. It's the current wonder concoction that's supposed to strengthen bones, heal wounds, lubricate joints, and improve immune function. In New York, long lines form in front of Brodo, a restaurant that specializes in bone broth customized with the likes of turmeric, chili oil, ginger, roasted organic garlic, shiitake mushroom tea, fermented beet juice, or bone marrow. If that doesn't please, you can choose an organic chicken broth, made with Pennsylvania Amish organic chicken, or a gingered grass-fed beef broth. It's reminiscent of Liebig's beef extract, originally developed in the 1840s as an inexpensive way to feed Europe's poor.

Liebig at the time was one of German chemistry's brightest lights, well-known for laying the foundations of the fertilizer industry by having determined the various nutrients plants need to grow. This whetted his appetite for the study of human nutrition; he hoped to boil down our needs to specific chemicals that could be supplied in a concentrated form. His idea was to boil bones in water and then evaporate most of the liquid to yield a thick molasses-like extract that became the prime product of Liebig's Extract of Meat Company. At first this was sold as a tonic to treat "weakness and digestive disorders," with claims becoming more extravagant as the popularity of the extract grew. It allayed "brain-excitement" if taken at night, claimed Liebig, and was also supposed to cure typhus and "inflamed ovaries." Hospitals, especially in England, made Liebig's extract a dietary mainstay, claiming it provided patients with the nutritive parts of animal foods in a remarkably concentrated form.

Popularity soon engendered skepticism that turned out to be warranted as analysis of the meat concentrate revealed that it was low in various nutrients and resulted in the death of dogs fed exclusively on it. That caused a change in marketing with emphasis being transferred from health to taste. It is still available today as Liebig Benelux, as well as in versions we know as Bovril and Oxo. Of course today's broth proponents would not dream of using a processed commercial product, instead focusing on cooking up bones with lots of vegetables. No question that can taste yummy. But as far as the health effects go? The claims are a bit hard to swallow.

Bone broth does contain collagen, which is indeed important for bone formation, joint function, and skin structure. But dietary collagen does not get transported to where it is needed in the body. Bone broth may taste very good, and may provide comfort when suffering from a cold, as does any hot beverage, but it is no magic elixir. Broth-promoting Chef Marco Canora, who owns Brodo as well as the adjacent Hearth restaurant, proclaims that proteins and amino acids in broth help with the lining of the intestines and that he himself has experienced a tummy benefit that has even alleviated his tendencies for depression. Liebig would be pleased. Modern chemists are skeptical.

They are also skeptical about the Nestlé company's launching of a novel instant coffee containing collagen that is supposed to enhance the appearance of the skin. The insinuation that the addition of collagen to coffee can have an effect on wrinkles amounts to nothing more than marketing puffery. Collagen is an important structural protein found in bones, ligaments, cartilage, and skin. It is a dynamic protein, meaning that it is continuously being produced and broken down. As is the case for any protein, the raw materials needed for its formation are amino acids. Cells called fibroblasts weave the individual amino acids into long protein chains. The required amino acids come from the diet, mostly from proteins we ingest.

During digestion these proteins are broken down into their component amino acids, which can then be used by cells to build the proteins the body needs. As we age, the production of collagen slows down, and the loss of collagen does become most noticeable in the skin, which becomes thin and wrinkled. This is where the seductive idea of supplementing the diet with collagen comes from. Since wrinkles are due to a loss of collagen, why not replace it by adding collagen to the diet? A rich idea for marketing, but it is poor in science. First, the reduction in collagen production with age is not due to a lack of amino acids in the diet. We eat plenty of protein to supply the needed amino acids. It is the chemical reactions that form collagen that slow down. Second, the idea that consuming collagen replenishes the collagen in the skin is sheer nonsense.

Like any other protein, during digestion collagen is broken down either into amino acids or into short chains of amino acids called peptides. These then go into the amino acid pool that the body draws on to synthesize the proteins it needs. Whether these amino acids originated from collagen or from soy protein is irrelevant. Third, even if dietary collagen could somehow replenish lost collagen in the skin, the amount added to a cup of instant coffee, 200 milligrams, is irrelevant in terms of the total collagen content of the skin. Chewing on a chicken wing or a pig knuckle would furnish far more collagen.

So far, Nestlé has only tried to pass off this silliness in Singapore. It seems that Asia is a more fertile ground for "beauty from within" products than the west. In Japan, collagen-fortified marshmallows are hot and restaurant diners can avail themselves of a meal made from soft-shell turtles to "boost their appearance." For men, the turtle meal is supposed to boost something else too. Probably has as much of an effect there as on the skin. Ladies can crunch on grilled meat frozen in blocks of gelatin to help beat wrinkles. This is hype without substance, but certain peptides derived from collagen are receiving attention as "nutricosmetics" by serious researchers. The notion is that at a dose of some 4 to 10 grams a day these can act as a false signal of the destruction of collagen in the body and trigger the synthesis of new collagen fibers that in turn can increase skin suppleness and reduce wrinkle formation. Needless to say, the wrinkles in this technology still have to be ironed out, but a Japanese company has already put the cart before the horse with Precious, a beer that contains 2 grams of collagen per serving and is marketed to women as a beauty treatment with the slogan: "Guys can tell if a girl is taking collagen." Unfortunately, it is doubtful that you can drink your way to beauty. Unless maybe you serve the beer to your partner. Beauty may be in the eyes of the beer holder.

Western marketers are beginning to catch on to the idea of eating for beauty as well. The French have come up with an anti-wrinkle jam containing fatty acids, lycopene, and vitamins E and C. And some restaurants in the U.S. are starting to offer "wrinkle-free meals." You can even get a cantaloupe extract that claims to protect the skin with its content of the antioxidant enzyme superoxide dismutase. More nonsense. Enzymes are proteins that like collagen are degraded during digestion. So, can anyone benefit from collagen-enhanced foods or beverages? Yes. The chicken industry, which is the source of the collagen that desperate baby boomers are swallowing to try to keep those wrinkles at bay. But what they are really swallowing with their collagen-fortified products is a good dose of hype.

TEA TIME

"Get in here and sit your ass down!" Not exactly what you expect to hear when you are peacefully walking in San Francisco's Chinatown. But the boisterous elderly Chinese gentleman seemed charismatic enough, and the establishment didn't look like an opium den. Indeed it wasn't. It was a teahouse. But not your ordinary teahouse.

My wife and I quickly found ourselves plunked down at a long counter along with a number of other tourists who had been dragged in from the street. "It's not for taste, it's for health," began "Uncle Gee," who I was to learn was a local institution. "Eighty-four years old," he boasted and "in perfect health!" "Drink eight cups of tea a day, never coffee!" "Tea full of antioxidants against cancer!" Not only were we treated to a lecture on the "science" and history of tea, we were also ordered to try about half a dozen varieties. "Never use boiling water. The tea will scream," and so did he. "Don't even dream about adding milk or sugar." "Steep for only twenty seconds!"

We sipped rosebud tea from Iran to ward off insomnia, and pu-erh for weight loss and heart problems. Next came Blue People Ginseng Oolong. I don't know why the "blue." I looked around and none of the people drinking it were turning blue. Everyone enjoyed that one. It was a truly different taste with a hint of licorice — "a party-in-your-mouth tea," we were told.

As our taste buds were partying, I glanced around at the dozens and dozens of jars, all with intriguing names. Monkey-picked green tea for "cleansing the body" caught my attention. It wasn't clear if this was to be applied to the outside or the inside of the body, or how the monkeys had been trained to pick tea. I wanted to ask if this was just monkey business but didn't dare. I did muster up enough courage to ask Uncle Gee about his favorite tea, the one that kept him young and so full of whatever. He quickly pointed to Angel Green tea. "Good for high blood pressure, cholesterol, diabetes, and detox." At $160 a pound, I suspect good for profits too, although there was no "hard sell." The tea bash ended with Uncle Gee telling us that while we were strangers when we came in, we were now part of the family. How could I resist buying some Angel and Blue People?

We've been enjoying both teas ever since, but other than frolicking taste buds, I can't vouch for any benefits. But tea leaves do contain over 700 compounds, many with potential biological activity. It is the polyphenols, the catechins in particular, that have aroused researchers' interest enough to generate a truckload of studies.

When rats are fed green tea leaves, their blood cholesterol and triglycerides go down. Levels of enzymes such as superoxide dismutase, catalase, and glutathione-S-transferase, all involved in removing foreign chemicals from the body, go up. The rats are also less prone to weight gain, apparently because of an increase in metabolism. But in these studies, the rats consume far more tea on a weight basis than people ever can. As far as human population studies go, some show a decrease in colon, breast, stomach, and prostate cancer, but others don't. The studies are neither consistent nor convincing, which is not surprising given that there are numerous varieties of tea with their chemical profiles depending on the type of tea, where it is grown, and how it is harvested, stored, and processed. Most of the epidemiological studies that have shown health benefits have focused on Asian populations where tea consumption is much greater than in North America and lifestyles are very different.

Laboratory studies have also been carried out with various

tea components. For example, heterocyclic aromatic amines, compounds produced when meat is cooked at a high temperature, are less likely to trigger cancer in the presence of the polyphenols theaflavin gallate and epigallocatechin gallate. Such findings, along with the suggestion of increased rates of metabolism, have led to the sale of various dietary supplements based on tea extracts. Why go to the trouble of drinking tea when you can just pop a "cancer-fighting, fat-burning" pill?

But here we run into a problem. Such dietary supplements are poorly regulated and the amount of catechins they contain can be far greater than that available from drinking tea. The high doses may indeed help to increase metabolism and result in weight loss, but the cost can be high. Just ask the teenager who walked into the emergency room at Texas Children's Hospital with his chest, face, and eyes bright yellow due to severe liver damage after using a concentrated green tea extract he bought at a "nutrition" store as a fat-burning supplement. There was concern that he may need a liver transplant, but luckily his liver, an organ that has regenerative properties, managed to recover. He did have to give up sporting activities and will require regular liver function checkups.

Unfortunately, this is not an isolated case, and such cases are not limited to green tea extracts. Recently, aegeline, a compound found in the leaves of the Asian bael tree, showed up in supplements marketed as aids to losing weight and building muscle, despite a lack of any credible evidence. But aegeline may not be without some effect. More than fifty people suffered liver damage: two had to have liver transplants, and one died after consuming a supplement containing aegeline. The multi-hospital Drug Induced Liver Injury Network in the U.S. has found that liver problems due to herbal and other dietary supplements have increased threefold in the last ten years.

Conventional medications still cause far more cases of liver injury, but they also have evidence of efficacy, which is not the case for many herbals.

I suspect that Uncle Gee would have a few devilish words to say about people who might think that they can encapsulate the benefits of Angel Green tea in a pill. He would likely argue that supplements could not replicate the same rejuvenating effects he experiences from his daily tea regimen. I must admit he did look robust and way younger than eighty-four. But when pushed, he did tell me that he runs six miles three times a week and can bench press 110 pounds. So maybe it's not only the tea that's keeping him young.

LITHIATED WATER

It was once called "The Texas Tranquilizer" because of its association with reduced admissions to mental hospitals and low crime rates. No, it wasn't a pill prescribed by physicians or a weapon wielded by law enforcement officers. It was naturally occurring ionic lithium in the water supply, particularly in the town of El Paso.

The theory about the calming effects of lithium on the population of the Texas town first emerged in 1971 when University of Texas biochemist Earl Dawson noted the presence of lithium in urine samples collected from some 3,000 citizens. He suggested the lithium must have come from the town's groundwater supply, which had a higher concentration of the element than is typically found elsewhere. Could this explain why Dallas with its surface water supply had seven times more admissions to state mental hospitals than El Paso? Could it also account for a crime rate that was half of that in Dallas, and a murder rate that was one twentieth?

There was already interest in lithium at the time because the U.S. Food and Drug Administration had just a year earlier approved the use of lithium salts for the treatment of manic illness. Although the idea that lithium could curb mania had been floating around since the late 1800s, it wasn't extensively embraced, possibly because this naturally occurring substance could not be patented and therefore was of little interest to pharmaceutical companies. But chitchat about the supposed benefits of lithium in water did send hopeful people scurrying to Lithia Springs, Georgia, to partake of its lithium-containing water. Luxury hotels mushroomed to welcome the rich and famous including Mark Twain, who is purported to have suffered from manic-depressive illness. But you didn't have to traipse all the way to Georgia to experience the legendary benefits of lithiated water. In 1887, a bottling plant was built, and the water was shipped around the country. Other marketers cashed in on the popularity of lithiated waters by just adding lithium bicarbonate to spring water.

Then in 1929, Charles Grigg decided to get a step up on the competition by adding citrus flavor and sugar along with lithium citrate to carbonated water. He called it Bib-Label Lithiated Lemon-Lime Soda and claimed that it would "take the 'ouch' out of 'grouch.'" The beverage was also a cure for hangovers, Grigg maintained. But the drink's name didn't exactly roll off the tongue, and he soon changed it to the simpler 7-Up. Why he chose the name isn't clear. Some suggest that it had seven ingredients and the "Up" referred to the mental lift it provided. Others claim the bottle contained 7 ounces and featured bubbles that rose when opened. Grigg took the secret of the name to his grave, but 7-Up is very much alive, although it no longer contains any lithium. The beverage was reformulated in 1950 after the FDA banned the use of lithium as an additive.

Water with naturally occurring lithium, however, can still be marketed. "Earth's Healing Magic in a Bottle" can be purchased from the Lithia Mineral Water Company, still located in historic Lithia Springs. Whether at 180 parts per billion (ppb) lithium has any biological activity is open to debate. This is way less than the dose used to treat mental illness, but in 2009 a Japanese study did link low levels of naturally occurring lithium in drinking water with an increased risk of suicide. Then two years later, the same group showed that even with the data adjusted for suicides, lithium exposure at levels even below 60 ppb was associated with a reduction in the standardized mortality ratio (SMR), albeit only by a few percent. The SMR is defined as the ratio of observed deaths to that expected in the general population.

The researchers then went on to raise a species of roundworm commonly used for antiaging studies in an environment where they were exposed to 60 ppb of lithium continuously and found that after twenty-five days, about 15 percent of the treated worms were still alive as compared with 10 percent of the untreated ones. Not exactly a stunning finding, but I guess if you are a roundworm, lithium might allow you to squiggle around for an extra day or so.

That study may just give producers of Happy Water, with its 100 ppb of lithium, sourced "from two ancient Canadian mountain springs," a little promotional wriggle room. It's doubtful that the water will put a spring in your step and a smile on your face, as the advertising suggests, but the claim that it is "free of empty calories" is good for a giggle. Contains full calories?

EATING BACON IS NOT THE SAME AS SMOKING

It's an all too familiar scene. A study is published linking some substance to cancer. It may be fluoride in toothpaste, butylated hydroxytoluene (BHT) in cereal, caramel in colas, oxybenzone in sunscreens, arsenic in rice, estrogens in milk, Para Red in paprika, aspartame in diet drinks, glyphosate in cereals, titanium dioxide in mozzarella, bisphenol A in canned foods, red dye #2 in maraschino cherries, tertiary-butylhydroquinone in chicken nuggets, nanoparticles in candies, chloropropanols in soy sauce, formaldehyde in pho noodle soup, dioxane in shampoos, chlorine in tap water, aflatoxins in peanut butter, PCBs in farmed salmon, microplastics in sea salt, pesticide residues on produce, acrylamide in French fries, or most recently, processed meats.

Before long scary headlines appear in the lay press: "Bacon Poses Same Cancer Risk as Cigarettes!" People panic and the industry attempts to soothe fears with arguments about poor quality research, cherry-picked data, and calculations about the gross amounts of food that would have to be eaten for the claimed effect to arise: "One would have to consume over 150 pounds of French fries every day in order to increase the risk of cancer from acrylamide," or, "just because we can measure something does not mean the levels are toxic; for farmed salmon, PCB levels were about 3 percent of the allowable limit of the Canadian Food Inspection Agency, the Food and Drug Administration, the World Health Organization (WHO), and the European Union."

Next, Dr. Oz hosts a segment on his show that makes a mountain out of a molehill, Gwyneth Paltrow tweets that she would rather smoke crack than eat the food containing the offending substance, the Food Babe organizes a petition to rid the world of it, and Stephen Colbert cleverly ridicules the risk. Devotees of the accused food unite to declare that they will not be deterred from eating it and highlight sketchy studies that show it is actually good for you: "Bacon is chock full of 'choline,' which helps increase our intelligence and memory and has been shown in university studies to help fight off the debilitating effects of Alzheimer's disease." Finally, more thoughtful media accounts begin to appear, adding perspective to the risk, often concluding that the answer to the chemical invasion is evasion through moderation.

Now let's process the meat of the matter. What do we make of headlines like "Processed Meats Rank alongside Smoking as Cancer Causes, Says the WHO"? WHOahh! No, eating a hot dog is not the same as smoking, and the World Health Organization did not say that it is. The current commotion was caused by a WHO advisory committee, the International Agency for Research on Cancer (IARC), officially classifying processed meats as being carcinogenic, placing them in the same category as tobacco smoke, asbestos, oral contraceptives, alcohol, sunshine, X-rays, polluted air, and inhaled sand. However, it is critical to understand that the classification is based on hazard as opposed to risk. Hazard can be defined as a potential source of harm or adverse health effect. Risk is the likelihood that exposure to a hazard causes harm or some adverse effect. If a substance is placed in IARC's Group 1, it means that there is strong evidence that the substance can cause cancer, but it says nothing about how likely it is to do so.

After examining some 800 peer-reviewed publications, IARC estimates that if a hundred people eat 50 grams of processed meat every day over a lifetime, one of them will develop colon cancer as a result. That is a small risk for an individual, but because there may be millions of people following such a diet, the impact on the population can be significant, estimated to be about 34,000 cases a year.

Consuming less than 50 grams of processed meat a day on average makes good sense, especially if it is replaced with plant products. But a hot dog, a salami sandwich, a few slices of prosciutto, and a couple of bacon and egg breakfasts a week are not suicidal. Two to three strips of bacon add up to 50 grams, as do two slices of ham, four slices of salami, or one hot dog. But if you have a smoked meat sandwich, alas, you've used up your weekly allotment. That may leave some people chewing their nails, at least until they read a headline about the male offspring of rats whose moms were fed phthalates, chemicals found in nail polish, having an unusually short distance between their anus and genitals.

NUTRITIONAL GUIDELINES — THEIRS AND MINE

There's a never-ending war being fought on bookshelves, websites, and in lecture rooms about what to eat to stay healthy, to lose weight, or thrive athletically. Actually, this isn't a new phenomenon. The ancient Egyptians believed that a diet rich in garlic increased energy, and Greek athletes spoke of eating mostly meat and eschewing bread before competitions. Shades of the Atkins diet. Regency-era poet Lord Byron explored all sorts of bizarre diets in a quest to look fashionably thin and pale. At one point he thought the secret was to be found in hard biscuits, soda water, and potatoes drenched in vinegar. Since he was the heartthrob of the era, many people followed his regimen. At the same time, formerly obese undertaker William Banting described how a program of four meals a day of meat, greens, fruits, and dry wine had finally worked for him. This was another forerunner of the modern low-carb diets. Banting avoided sugar, starch, beer, milk, and butter and gave rise to the expression "to bant" among his followers. Meanwhile over in Germany, chemist Justus von Liebig advocated eating meat and its juices for maximal nutrition and even founded Liebig's Extract of Meat Company to produce an inexpensive nutrition source for Europe's poor. (For more on Liebig, see "Boning Up on Collagen.")

Today we have dietary schemes galore. There's the blood type diet, the alkaline foods diet, the food combining diet, the intermittent fasting diet, the Inuit diet, the Okinawa diet, the low glycemic index diet, the 100-mile diet, the macrobiotic diet, the Mediterranean diet, the dash diet, the Beverly Hills diet, the fruitarian diet, the master cleanse diet, and the tapeworm diet. We even have scientifically challenged bloggers who believe that the safety of a substance is determined by the number of syllables in its chemical name. The proponents of these schemes may be physicians, scientists, celebrities, or scientific nobodies who are often adept at torturing data until it submits to their pet theory. Controversy abounds.

What we really need are some straightforward evidence-based guidelines that are easy to follow. The U.S. government tries to offer such every five years with its *Dietary Guidelines for Americans*. The most recent ones, just released, undoubtedly produced elation among egg, coffee, and oil producers, dismay in the sugar industry, sighs of relief in the meat and soft drink sectors, anger among some nutritional scientists who had served on the government's advisory committee and, as usual, confusion among the public.

Here are the nuts and bolts. Dietary cholesterol is no longer a "nutrient of concern," liberating eggs from their shackles. Somewhat oddly, though, the guidelines also state that people should eat as little dietary cholesterol as possible. Three to five cups of coffee a day can be part of a healthy diet, and as far as fat goes, only the saturated variety is to be limited to 10 percent of total calories. Otherwise there are no limits on fat consumption, to the delight of the vegetable oil producers. For the first time, sugar is singled out as a problem, with advice to limit it to 10 percent of total calories.

What is the problem here? While there is specific advice about what foods to eat, with beans, peas, different colored vegetables, grains, nuts, and yogurt being elaborately described, when it comes to foods to limit, specifics are replaced by nutrients. Cut down on sugar is the advice without mentioning that a very effective way to do this is to cut down on pop. Just one sugar-sweetened soda exceeds the recommended added sugar intake for a day. There is no mention of health concerns that have been raised about eating red meat other than urging men and boys, who according to surveys eat too much protein, to "reduce overall intake of protein foods by decreasing intakes of meat, poultry, and eggs." Cutting down on red meat consumption would also help in meeting the 10 percent saturated fat goal, although it should be said that whether saturated fat is indeed linked to heart disease is hotly debated. Furthermore, reducing consumption of foods with saturated fats may offer no benefit if they are replaced by pasta made with refined carbohydrates, as is often the case. The advisory committee had suggested that these specifics be included, even urging that the environmental benefits of reducing meat consumption be mentioned, but apparently the meat and beverage industries were successful in whispering sweet nothings into the right ears to prevent their products from being singled out.

I have no vested interests, so I have no problem mentioning specifics, as you can see from my attempt to distill guidelines out of the plethora of published nutritional information. Here we go. Eat steel-cut oats mixed with oat bran and topped with a spoonful of flaxseed and berries for breakfast. Red meat no more than once a week. Processed meats no more than once a week. Pastries no more than twice a week. Fish a couple of times a week. Fruit and nuts for snacks and desserts. Whole-grain bread instead of white, olive oil for salads and cooking, vegetables with every meal. Focus on colors other than white on the plate. Don't worry about butter on your bread or consuming eggs, coffee, or tea. "Gluten-free" is relevant only if you have celiac disease. "GMO-free" is irrelevant. An alcoholic beverage a day is fine. No need to force milk. Avoid trans fats; they are there if you see "hydrogenated" on the label. Soft drinks are a pariah. At night, count the servings of fruits and veggies you had all day, and if less than five, search the fridge for apples. Cook rather than buy. There is nothing that you should never eat, but if you are not hungry, don't eat. And if you stray from this plan, don't worry, but don't make a habit of it. But do make a habit of exercising. It probably matters more than what you eat.

JEANS TO PURIFY AIR

You may want them in your jeans, but you probably want to keep them away from your genes. They're nanoparticles of titanium dioxide, about ten billionths of a meter in diameter, that can exhibit beneficial properties not possessed by their larger cousins, but they may also have a darker side.

There are more jeans in the world than people. That stat sparked an idea in the mind of University of Sheffield chemist Tony Ryan. Why not use people's penchant for wearing denim to help purify the air? After all, the International Agency for

Research on Cancer (IARC) classifies outdoor air pollution in Group 1, reserved for substances that are known to cause cancer in humans. It estimates that there are up to seven million premature deaths in the world every year as a result of air pollution. With thoughts of reducing pollutants such as the nitrogen oxides and volatile organic compounds emitted by vehicles, power plants, residential heating, cooking, and various consumer products, Ryan, in partnership with former fashion designer Helen Storey, came up with the concept of "Catalytic Clothing."

Catalytic apparel uses fabric impregnated with nanoparticles of titanium dioxide to degrade air pollutants. "Nano" means small. So small that the combined surface area of the nanoparticles that are distributed through any fabric is immense. And that matters because the action takes place on the surface of the particles.

Titanium dioxide is a photocatalyst, meaning that it can make chemical reactions happen when exposed to the right wavelength of light, in this case ultraviolet. The light energy causes it to release electrons that then target water molecules in the air, breaking them apart to form extremely reactive hydroxyl radicals that then chop up organic compounds into simple molecules such as carbon dioxide and convert nitrogen oxides into water-soluble nitric acid. This is not just theory, it is well-established technology that already has commercial application, for example in self-cleaning glass. A thin layer of titanium dioxide ends window cleaning worries, as long as the climate provides for sufficient sunshine and rain. The chemical can even be mixed into concrete, resulting in self-cleaning buildings such as the Jubilee Church in Rome.

Thanks to titanium dioxide, we may never have to confront yellow urinals again. Coating the ceramic with a layer of titanium dioxide, about a fiftieth the thickness of human hair, prevents stains from forming. The technology also has potential in operating rooms, where bacteria on floor and wall tiles can be destroyed with fluorescent light, common in hospitals, furnishing enough of the right wavelengths. And how about self-cleaning tiles for the kitchen and bathroom?

Clearly, titanium dioxide photocatalysis is sound technology. But can wearing jeans treated with this chemical actually have an impact on air pollution? According to Professor Ryan, yes. He calculates that that if a third of a million people in Sheffield wore such jeans, nitrogen oxide levels could be significantly reduced. And there is no need to buy special jeans. Titanium dioxide particles stick readily to the fabric, so the idea is to add a formulation of the chemical to the water when the jeans are being laundered. The nanoparticles will stick until the fabric degrades.

As is often the case in science, there is a "but." What happens if nanoparticles enter the bloodstream? What tissues might they affect? Titanium dioxide has the potential to damage DNA, but to do that it has to enter cells. That is a possibility since nanoparticles are smaller than cells. In the lab, nano-titanium dioxide has been shown to damage DNA in human intestinal cells, but only at doses far higher than what could ever be ingested.

In any case, people will not be dining on their treated jeans. But they may be gulping donuts, or a vast array of other foods such as Gobstoppers, M&M's, pastries, or soy milk that have titanium dioxide added to them to provide a more pleasing whitened appearance. Only about 5 percent of the titanium dioxide is made of nano-sized particles, but that has raised concern because the IARC has classified titanium dioxide as possibly carcinogenic to humans (Group 2B). This classification is based on inhalation of titanium dioxide dust in an occupational setting, quite a different exposure than eating a donut with a titanium dioxide–enhanced white sugar coating. Nevertheless consumer activism has resulted in Dunkin' Donuts removing titanium dioxide from the powdered sugar coating on its products. Maybe it can be redirected into catalyst jeans. We really don't need to make junk food look more appealing, do we?

AMAZING CHARCOAL

It's a killer. It's a savior. It's also a trickster. It's one of the most important substances ever discovered. It's charcoal! Burn any animal or vegetable matter with a limited supply of air, as is the case inside a woodpile, and you are left with charcoal, essentially carbon mixed with some mineral ash. The fact that charcoal burns better than wood was probably noted soon after man learned to control fire over a million years ago. The first use of charcoal for purposes other than providing heat was around 30,000 B.C. when cavemen used it as a pigment for drawing on the walls of caves.

Then around 4000 B.C. came a monumental discovery, probably by accident, when a piece of ore fell into a charcoal fire and began to ooze metal. When naturally occurring ores of copper, zinc, and tin oxides are heated with charcoal, the carbon strips away the oxygen, leaving the pure metal behind. Alloying copper with tin forms bronze. The Bronze Age was followed by the Iron Age, characterized by the smelting of iron from iron oxide with charcoal. That same technology is still used today, so charcoal literally continues to shape our world. But it wasn't only through the smelting of metals that charcoal impacted history.

Sometime in the ninth century, a Chinese alchemist discovered that blending charcoal with saltpeter (potassium nitrate) and sulfur resulted in a mixture that would combust readily.

Gunpowder would eventually be used to create explosives that gave access to coal and minerals, making huge engineering achievements possible. Of course gunpowder also made possible the easier destruction of life, casting a dark shadow on charcoal.

Around 1500 B.C. Egyptian papyri recorded the use of charcoal to eliminate bad smells from wounds, the first mention of a medical application of charcoal. By 400 B.C., the Phoenicians were storing water in charred barrels on trading ships to improve its taste. It seems they had hit upon one of charcoal's most important properties, the ability to bind substances to its surface, a phenomenon known as adsorption. That application lay more or less dormant until the late eighteenth century, when Europeans developed a taste for sugar. Raw sugar from sugarcane or sugar beets is tainted by colored impurities that can be removed by passing sugar extract through beds of charcoal.

The rapid growth of the sugar refining industry led to a search for charcoal with improved adsorption properties and resulted in the development of activated charcoal, also referred to as activated carbon. In this process, carbonaceous matter such as wood, coal, or nutshells is first heated in the absence of air, followed by exposure to carbon dioxide, oxygen, or steam. This has the effect of increasing the surface area and establishing a network of submicroscopic pores where adsorption takes place. Later, it was determined that impregnation with chemicals such as zinc chloride or phosphoric acid prior to heating improved the adsorption properties. Today a variety of activated carbon products are available for use in various applications. And there are applications galore, ranging from the medically and environmentally important to the frivolous.

Activated charcoal is used in water filters, air purification systems, gas masks, and even underwear. Yes, flatulence filtering undergarments for people suffering from various gastric problems really works. But in order to avoid flatulence escaping around the filter, the patient is recommended to stand with legs together and let the wind out slowly.

Because of its amazing adsorptive properties, activated carbon is a staple in emergency rooms. In cases of suspected drug overdose or poisoning, it is administered orally to bind the toxins before they have a chance to be absorbed into the bloodstream. It isn't surprising that inventive marketers have absorbed this information and have started to roll out various foods and beverages containing activated carbon with promises of "detoxing." Black Magic Activated Charcoal, a "zesty lemon detox and purification elixir," invites you to "come over to the dark side." A very apropos invitation. Just what sort of toxins is this beverage supposed to remove? And since activated carbon isn't very specific in what it adsorbs, it is as likely to remove vitamins, polyphenols, and medications as those unnamed toxins. Of course it is made with "alkaline water," catering to the nonsense that cancer is caused by an acidic pH. Any alkaline water is of course immediately neutralized by stomach acid. Believe it or not, you can also get activated carbon ramen noodles. The only thing these will eliminate is your appetite.

A FASHIONABLE ADDRESS

The most fashionable address in New York is no longer Central Park West. It is 66 East 11th Street in Greenwich Village. The building doesn't look like much from the outside; it was once a factory then a parking garage. Now totally remodeled on the inside, it has been dubbed as "wellness real estate" with a focus on the environment and the health of the residents. A healthy wallet is definitely a requirement for moving into a home that claims to support cardiovascular, respiratory, and immune health through a variety of amenities that range from showers infused with vitamin C to photocatalytic coatings on surfaces designed to counter contamination by microbes. Prices of units range from about fourteen to fifty million dollars. But for that you get to hobnob with neighbors such as Leonardo DiCaprio, which may be a plus, or Deepak Chopra, a definite minus.

Chopra is the New Age guru who has amassed a fortune with his confused and confusing books in which he rambles on nonsensically about how "we are thoughts that have learned how to create the physical machine, the body" and that "there is no physical world, it's all projection." He adds that "the whole thing is a Quantum Soup and reality exists because you agree to it." Well, there is a reality Chopra seems to have agreed to. And that is to shill for Delos, the "pioneer of Wellness Real Estate," the company that has designed his multimillion dollar abode. The stunning habitat comes with a grab bag of science and nonsense.

There is a special water filtration system that "reduces disinfectant by-products, chlorine, pesticides, and some pharmaceutical and personal care products." Nothing unusual here, many such systems based on activated carbon and ion-exchange resins are available for home use. The question then is why there is a need for "shower water infused with vitamin C, which neutralizes chlorine to promote healthy hair and skin." Hasn't the filter system already removed the chlorine? Yes it has, otherwise it would be a pretty useless system. So the only reason for the vitamin C infusion is to get some mileage out of the common association of vitamin C with health. Vitamin C can indeed neutralize hypochlorous acid, which is the active form of chlorine in water, but it does not do so very efficiently. A gram of vitamin C would eliminate chlorine from about 400 liters of water, which is roughly equivalent to three to four showers. And there is no evidence that this would have any effect on hair or skin.

Installation of an air purification system that filters pollen and other small particulate matter and uses ultraviolet light to kill microbes in the air ducts does get marks. But there is less scientific support for incorporating substances, in all likelihood titanium dioxide, into counters and floors to destroy bacteria on contact. This may be welcome in an operating theater, but there is no need for such antibacterial warfare in a home. Disease-causing bacteria do not lurk around every corner, and we happily coexist with the vast majority of bacteria. But generally bacteria are regarded as public enemies, and antibacterial claims are good for marketing. Rinsing surfaces with soap and water serves us just fine.

Much is made of outfitting the rooms with just the right kind of lighting to prevent the body's biological clock from going off-kilter. Here we do have some science. Light of any kind suppresses the secretion of melatonin, the so-called "Dracula hormone," but blue wavelengths do it more effectively. Melatonin is produced during darkness and is associated with sleep. In the morning when we want to boost alertness, suppression of melatonin production is desirable. This is why Delos installed lights with a blue emphasis in showers and around bathroom mirrors. Hopefully people who like to take their showers at night can switch off these lights.

During the day, emphasis should be on wavelengths other than blue to prevent a big drop in melatonin, which has been associated with adverse health effects. Studies have linked working the night shift and exposure to bright light to cancer, diabetes, heart disease, and obesity, possibly due to low levels of melatonin. There have been suggestions that night workers wear glasses that filter out blue wavelengths in order to boost their melatonin levels. Might be a good idea to have such filters on reading lamps that are used before falling asleep.

For improved sleep, total darkness is desirable to enhance melatonin production, and accordingly Delos has equipped windows with programmable blackout shades. Night lights are red since these wavelengths have the least power to suppress melatonin and shift circadian rhythms. Delos's advertising correctly explains that lighting can affect health, but then goes on to say that protection is needed from electromagnetic fields that disrupt sleep. There is no evidence that EMF fields are harmful nor that "electromagnetic field panels" are of any use.

Speaking of questionable benefits, is there data to back up claims that "impact absorbent floors improve lumbar support"? Noise reduction with soundproof sheetrock sounds great, but the claim that noise decreases the production of telomerase, an enzyme associated with youth, is at best speculative. And how about the "reflexology path" in the bathroom, featuring an uneven floor with hard protrusions to stimulate acupressure points that are supposed to stimulate energy meridians in the body? Not exactly hard science.

If you are lacking the millions to purchase a Delos condo, you can still experience the "wellness" effects with a stay at the MGM Grand in Las Vegas. A number of rooms have been outfitted with the same "health" amenities, including a special TV channel where holistic guru Deepak Chopra greets guests and offers advice about using acupuncture instead of Prozac and eating pink food for fewer wrinkles. That's a reference to astaxanthin, a carotenoid that is responsible for the pink color of salmon. Astaxanthin may actually offer some protection against sun-induced skin damage, but only when taken in supplement form.

Chopra goes on to inform the lucky guests that they will

be experiencing "the next frontier in well-being" and an environment that "basically allows your body to self-regulate." I think just the prospect of turning on the TV and possibly seeing Chopra mutter about "quantum consciousness" would keep me from forking out the surcharge for a wellness room at the MGM. It would make me feel unwell.

FINAL THOUGHTS

Most chemistry conferences these days feature a session on the "public understanding of chemistry." Usually speakers express frustration about equating the term "chemical" with "toxin" or "poison," about consumers looking for "chemical-free" products, and about the extent of scientific illiteracy. There tends to be a collective bemoaning of the lack of appreciation of the contributions that chemistry has made to life and of the eyebrows raised when a chemist reveals his profession in some social setting. Annoyance surfaces about synthetic chemicals being seen as the culprits responsible for a host of human ailments whereas natural substances are judged to be unquestionably safe.

Often there is criticism of bloggers who maintain that if you can't pronounce the chemical name of a food ingredient you shouldn't be eating it and of the bothersome image of the frizzy-haired "mad scientist" who is bent on brewing up some nasty carcinogen to unleash on an unsuspecting public. There's lots of lamenting the demonization of "petroleum-derived" chemicals by scientifically uneducated, self-appointed protectors of the public good.

Speaker after speaker expresses concern that the public is being unduly alarmed by ill-informed pundits who inflate the risks of nonstick cookware, fluoride, pesticide residues, preservatives, plasticizers, GMOs, and various chemicals found in cosmetics and cleaning agents. There is also concern that chemists are unfairly maligned, mistrusted, and uncaring about the long-term consequences of their actions. All of this is usually followed by a call to arms to change the public's attitude toward chemistry, and vigorous discussions ensue about how to go about curing what is seen as widespread "chemophobia." I know, because I've been there and have taken an active role in such dialogues.

Now, though, it seems that our worries may have been overblown, at least judging by the largest survey ever carried out about the public's attitude toward chemistry by the U.K.'s Royal Society of Chemistry. A qualifier has to be mentioned here though. In the U.K., pharmacists are also called chemists, and this likely skewed the statistics since health professionals tend to be regarded in a positive fashion.

The survey featured interviews with over 2,000 randomly selected people and discussions with a number of focus groups. While there were concerns about chemicals, chemistry as a profession was viewed positively. Sixty percent of the subjects interviewed said they believed that the benefits of chemistry outweigh any harmful effect, and 84 percent agreed that chemists make a valuable contribution to society. Interestingly, only 12 percent of chemists interviewed thought the public would have such a high appraisal of their profession.

When it comes to chemicals, 70 percent agreed that everything can be toxic at a certain dose, but only 60 percent knew that everything is made of chemicals. On the positive side, less than 20 percent thought that all chemicals are dangerous. So chemophobia does not seem to be as extensive as we think it is.

Chemists have a knee-jerk reaction every time we see the

word "chemical" used in what we consider to be an inappropriate fashion. We bristle when someone says they do not want to eat food that contains chemicals or when we hear that consumers are looking for a cleaning agent without chemicals. What ignorance, we think! But it seems that when people use "chemical" in this fashion, they are referring to substances that they believe are potentially toxic, not to all chemicals in general. It's a matter of semantics. Maybe we are wasting our time by trying to set the record straight every time we see the word chemical used in a way that strays from our scientific definition. Perhaps it is time to accept that words can have different meanings depending on their context, and that when laypeople talk about "chemicals" they are using the term to mean substances that are potentially harmful.

Ridiculing the misuse of the word as a synonym for "toxic," as those of us in the chemistry field often tend to do, can have an undesired consequence. It can give the impression that we think that all chemicals are safe. In fact, no one knows the potential harm that can be caused by some chemicals better than chemists.

An unreasonable attack mounted against some chemical by a chemically illiterate person is sometimes interpreted by chemists as an attack on their profession and prompts a vigorous rebuttal. Even if scientifically warranted, it tends to project an image of chemists being defenders of all chemicals.

As scientists, chemists are gung ho on evidence and are wary of anecdote. Yet, it appears that our belief that chemists are considered societal pariahs because they produce chemicals, that is, "toxins," is purely anecdotal. The U.K. survey actually revealed that 75 percent of people think that chemistry has a positive impact on our well-being.

Admittedly, I was surprised by that statistic, probably having been misled by my personal anecdotal evidence. Because of the business I'm in, I tend to take note of any chemical nonsense I come across. I see it in my emails and on posts on my Facebook page. And I guess I forget that the vast majority of people who have a reasonable view of chemicals and chemists are not vocal about their beliefs. It's the squeaky wheel that we hear.

Thanks to the Royal Society of Chemistry's survey, we can now move from anecdote to science. It is comforting to note that chemophobia is not rampant and that only 25 percent of people are confused, bored, shocked, saddened, or angered by chemistry. But there is another noteworthy statistic. More than half the people do not know what chemists actually do and do not feel confident enough to talk about chemistry. So instead of worrying about the misuse of the word "chemical," we should focus on educating the public about the role of chemistry in our lives. I hope that with this book I have managed to do some of that and have whetted your appetite for further feasting on the delicacies of science.

INDEX